D0391935

A NEW KIND OF BOOK FOR A NEW ERA

With *First Light of Day,* Mike Steep has accomplished something extraordinary. Not only has he written an incredible description—chilling as it is—of the world ruled by machine intelligence that is just around the corner, but he has done so by inventing a new kind of book for this new era. It is a hybrid of a novel and a nonfiction book, each part furthering the vision of the other. It is as revolutionary as the future it explains.

—**Michael S. Malone,** author, columnist, editor, investor, businessman, and television producer

PRAISE FOR *FIRST LIGHT OF DAY*

What is the price of our society's obsession with artificial intelligence? Michael Steep's First Light of Day *paints a picture of the not-too-distant future, illuminating the dangers of a world of our own creation, but not of our own design. In so doing, Steep's lucid and heartful narrative makes an impassioned case for each of us to remember our humanity, especially in the digital age.*

—**General Stanley McChrystal,** partner, McChrystal Group

An expansive trip into our future and the technology to take us there. Michael Steep has created a new genre of technological realism that gives the reader a heady combination of science fiction and scientific fact.

—**Tom Wheeler**, former chairman, Federal Communications Commission; author, *From Gutenberg to Google: The History of Our Future*

Mike Steep gives a gripping look at how artificial intelligence and digital technologies will fundamentally change the future for every individual and organization. In his fictional narrative, he provides a fascinating look at life in the near future based on his treasure trove of experience of more than three decades at the world's most admired technology companies.

—**Michael Morell**, former deputy director, CIA

Using his Silicon Valley real-world experience, Michael Steep has created a novel exploring big ethical dilemmas as AI surpasses human capabilities. AI is either humanity's saving or its demise. First Light of Day, *including its bonus explanation of key-ingredient technologies, gives the reader a front-row seat to a possible future where nationalism, politics, and our essence as humans confront rapid change in technology.*

—**Gary Shapiro**, president and CEO,
Consumer Technology Association

Michael Steep has written an engaging and insightful novel that explores how multiple personal AI assistants like Siri and Alexa, which many of us already use today, are inexorably advancing in their capabilities to the point where they can ultimately develop "personalities" and "self-awareness."

Steep has developed a wide-ranging set of insights about digital and other emerging technologies through a long career in commercializing solutions for tech companies like HP, Apple, Microsoft, and Xerox PARC, and as the executive director of Stanford's Disruptive Technologies and Digital Cities research program. The book masterfully weaves these insights about the trajectory of these disruptive technologies into a plausible and frightening story that shows how these intelligent assistants coevolving and interacting in unexpected ways can create a dark and dystopian future of severe economic inequality, cybercrime, and conflict.

The book is an important and page-turning read for anyone working on, using, and/or concerned about the advances of intrusive Big Data collection and the artificial intelligence applications that feed ravenously on it.

—**Raymond Levitt,** operating partner, Blackhorn Ventures, LLC;
Kumagai Professor of Engineering, Emeritus, Stanford University

Is it science fiction, or our future? Steep has opened up the integrated experiences, business models, and societal impact that disruptive technology portends. As a country, we have big choices to make, and First Light of Day *tells a captivating story highlighting the powerful changes already underway. As a former leader at Amazon, I remember that, when the company started 25 years ago, how ridiculously traditional business leaders underestimated the impact of Amazon and digital technologies. Now, looking forward 10 to 20 years, what sounds the impossible will become the everyday. Enjoy the story and don't ignore its lessons!*

—**John Rossman**, managing partner, Rossman Partners; author, *Think Like Amazon: 50½ Ideas to Become a Digital Leader*

I've known Mike Steep since he hired my company 20 years ago to do PR for his startup Mira Technologies. We're both card-carrying disrupters, having lived through numerous paradigm shifts driven by Silicon Valley. Challenging people to think is in his bones.

Few would attempt what he's created with First Light of Day*—it propels us forward 35 years to experience a totally plausible and very challenging future built on the actual technologies we are deploying today. He dares us to look at what we have all been a part of creating and to think about what the future will be like. This book educates as well as entertains, with a Part Two that explains the technologies that underpin the fictional Part One.*

—**Andy Cunningham**, former PR executive for Steve Jobs, Apple Computer; orchestrated the launch of the Apple Macintosh in 1984; founder and CEO, Cunningham Collective

I have known Mike Steep since his time at Apple. He has consistently looked into the future and identified those technologies that would go on to impact everyday life. In this fast-paced mystery, Steep has crafted an intriguing story about how technology will forever change the way we live.

—**Don Strickland**, president and CEO, Strickland Associates; adjunct professor, Imperial College London

Michael Steep's First Light of Day *provides our team a priceless look into not just the future, but the ramp from the present to the future. Diamond Wealth is a boutique financial-services firm and a member of Stanford's disruptive technology program. Our business depends on information and insights. This book helps us understand how disruptive technologies can and likely will impact society.*

—**Ronald S. Diamond**, chairman and CEO, Diamond Wealth Strategies, Family Office

FIRST LIGHT OF DAY

FIRST LIGHT OF DAY

A CAUTIONARY TALE OF OUR FUTURE WRITTEN BY ONE OF TODAY'S LEADING EXPERTS ON TECHNOLOGY INNOVATION

MICHAEL J.T. STEEP

Founder and Executive Director,
Disruptive Technology and Digital Cities Program
Stanford School of Engineering

With Dr. Herman Donner, Stanford School of Engineering

First published in 2020 by Silicon Valley Press.

ISBN: 978-1-7339591-0-0
ISBN: 978-1-7339591-1-7 (ebook)

Library of Congress Control Number: 2019910757

DEDICATION

This book is dedicated to the memory and life of my dear friend Mark French, founder and CEO of Leading Authorities.

> "My advice to my friends is to take nothing for granted. It can all change very quickly. Savor the time with family and friends, and invest your time and emotional energy with those you love. Add to this: be kind to others."

This book is also dedicated to the memory of Robert Lucid (1930–2006), professor of English at the University of Pennsylvania; and Norman Mailer (1923–2007), novelist, journalist, political activist, filmmaker, and guest teacher. I was a freshman student in Robert Lucid's class "Art of the Novel" when I committed to publish my first novel. I have now fulfilled that promise both to Professor Lucid and myself.

Finally, I dedicate this book to our future generations—including Miguel, Aryana, Mari, Diego, Jules, and James—who will have to manage through a new social world order.

CONTENTS

FOREWORD

The book you are about to read not only describes the complicated, chaotic, potentially threatening world that awaits us but also *portends* that future. In its design, it is a chimera, an unlikely yet wonderful hybrid between a science-fiction thriller and scientific paper.

When I first opened *First Light of Day,* I was skeptical. How could a book like this possibly work? But as I read its compelling narrative, I understood—this is the only way one can approach such a complex subject.

The new world hurtling toward us, full of artificial intelligence, thinking machines, robotics, and the Internet of Things, promises to be so revolutionary and disorienting that perhaps the only way we can conceptualize it is through fiction. No nonfiction description could possibly capture the human dimension of this metamorphosis—the emotional, internal explosion about to take place in each of our lives. I should know; I spend an enormous amount of time trying to get these points across.

Yet to understand *how* today's world will be upended by the penetration of smart tech into every corner of daily life, the reader also needs an in-depth briefing on the technology itself. This demands a different type of narrative: precise, objective, and unsentimental.

How, then, to combine the two? How to tell a compelling story, while underpinning it with empirical information to make that story not only believable but also *possible*?

The novel *First Light of Day* is a dramatic, exciting, and chilling work of art. This is the dystopian, Black Mirror–style future that we who work in technology fear most, the one that awakens us, sweating, in the night. Mike Steep has made this story even more powerful by driving the plot toward what we all sense is *the* turning point in human history: the first fully conscious computer. Steep tackles this challenge in the most clear-eyed way imaginable. It is fearless writing of the highest order. You are in for a (terrifying) treat.

Then, when you finish the novel, rattled and telling yourself that this is "just fiction," Steep presents you with a full exegesis of all the technology that he's just described. Yes, he's telling you, this scenario can, indeed, happen—that, in fact, it's the *most likely* scenario for the future. If you still have doubts, he presents real-life examples of where the first steps toward this new world have already occurred.

Mike Steep is one of the few people who could write such a memorable and innovative book. As executive director of the Disruptive Technology and Digital Cities Program at Stanford University's Global Project Center, he is one of the world's leading expert on the impacts of the new communications, transportation, and information technologies on modern urban life. When he describes the future through the lives of Mikhail, his friends, lovers, and enemies in the London of 2050, the depiction is likely to be more accurate than any other writing today. There are many experts on this topic, but few with Steep's credentials.

What makes Steep unique is that he is also a trained novelist, having been taught by the legendary writing professor Bob Lucid. A hugely successful career in corporate America—including tenures as a senior executive at Apple, Microsoft, and Xerox PARC—may have lured him from a career in literature, but he never gave up his love of fiction writing.

With *First Light of Day*, Mike Steep at last fulfills his early literary promise. This book will entertain you, stun you, and shake you to the core. When you finish reading it, you will have traversed a remarkable creative achievement. More importantly, you will have experienced something surpassingly rare: an early glimpse into our common fate.

—Salim Ismail, Singularity University

PART ONE

THE NOVEL

INTRODUCTION TO PART ONE

In April 1968, Stanley Kubrick's masterpiece, *2001: A Space Odyssey,* premiered at the Loew's Capitol Theatre in New York City, the same theater where classic movies, including *Gone with the Wind* and *The Wizard of Oz,* had opened in the 1930s. I was a boy of 15 at the time, on a school field trip to New York City. I could never have imagined that a science-fiction thriller would have the impact on my life that this movie did.

As I left the theater, I was determined to create the mythical and omniscient computer in the film, the HAL 9000 series, a quest that ultimately would lead me to Silicon Valley. That decision, which would ultimately result in a 40-year career working with companies such as Hewlett-Packard, Software Publishing Corporation, Apple, Microsoft, and Xerox PARC, led me to where I am today.

Although I never did achieve my quest to create HAL, I did participate in the launch of the first minicomputers, the personal computer, and Microsoft's Azure Cloud Service. My team licensed Software Publishing's PFS Series business suite to serve as the software for the launch of the IBM PC. In the 1990s, while at Apple Computer, we introduced the world's first consumer digital camera (called the QuickTake). We also developed the Apple LaserWriter as part of the desktop publishing revolution. At Microsoft, I witnessed the rise of the mobile phone, managed the transition from packaged software to software-as-a-service, and worked with the Azure team on its first cloud offering as part of the Enterprise Partner Group. At PARC, we developed and licensed early-stage technologies crossing advanced material sciences, predictive data analytics, clean energy, and behavioral software. We witnessed the introduction of the BMW iSeries electric car and the first electric airplane flown from England to France.

Now, as the cofounder and executive director of the Disruptive Technology and Digital Cities Program at Stanford Engineering School, I realize that the enormous growth in technology during my career was simply the opening act to an even more dramatic play. We are now in the midst of a monumental global change in scientific invention and innovation. This exponential growth in disruptive technologies is greater than anything that has come before, outpacing our ability to manage it and changing the way we conduct our business and personal lives. Pandora's box has been opened for the human race.

All these advances in technology—data analytics, computing power, cloud, networking, and a host of other technologies—are converging to create a perfect storm. Every major commercial company in every global industry now faces unrelenting disruption of their business models. At the same time, the human ability to understand and manage the integration of disruptive technology is rapidly falling behind. This massive gap between exponential growth and our linear human ability to cope widens by the year. As a result, humans have, for all intents and purposes, lost control of the pace and impact of technological change.

Part One of the *First Light of Day* is a novel written to shed light on—and offer a warning about—how current-day technological advancement will define the world our children and grandchildren will likely face. Part Two, a nonfiction companion guide to the novel, explains where we are in the development and commercial launch of each of these disruptive technologies. You can enjoy the fictional part of the book alone or read Part Two to attain a deeper technical understanding of the events presented in the novel. Part Two also includes suggested questions for discussion, as I do hope this book inspires people to question what is happening and what can happen, in the not-so-distant future.

Another thought for the reader to consider: just as we are experiencing an exponential explosion in technology, we are also facing other potentially cataclysmic changes, including global warming and the political-economic upheaval of globalization. We are at a turning point for the human race, perhaps more dramatic than at any other time in human history. In his address to the 2018 Climate Change Conference in San Francisco, Thomas Friedman noted that the "simultaneous accelerations in the Market, Mother Nature, and Moore's Law together constitute the 'age of accelerations,' in which we now find ourselves."

Technology can be both a blessing and a threat. But the one thing I would encourage us all to do is to pause and consider the unintended consequences of each of our actions. The power of disruptive technology is now unleashed, but we still have the ability to direct how it will be used, for better or worse. The ultimate question we all need to answer is: What path will we choose?

PROLOGUE

England, you had better go,
There is nothing else that you ought to do,
You lump of survival value, you are too slow.

England, you have been here too long,
And the songs you sing are the songs you sung
On a braver day. Now they are wrong.

And as you sing the sliver slips from your lips,
And the governing garment sits ridiculously on your hips.
It is a pity that you are still too cunning to make slips.

—from "Voices Against England
in the Night" by Stevie Smith

LONDON, APRIL 2045

The Fifteenth Convocation of the Modern World assembled under the auspices and blessing of the Bishop of St. Paul's Cathedral, and the direction of the Lord Mayor of the City of London. First built in 604 AD, the cathedral had become one of the bedrock symbols of the transition of the British system post-Brexit into the era of the High Value Citizen. On this day, 1,500 émigrés congregated for their formal induction into the *Arrivés* of London society. Like the privileged aristocrats of a bygone era, the *Arrivés* were anointed by English society as the aristocracy of the modern era.

British society had changed significantly in the previous decades, splintering into two distinct classes of citizens: those who had contract employment and those who were either underemployed or permanently unemployed. At the very top of the contract employment were these *Arrivés*, or High Value Citizens, who provide the valued technology services to drive economic growth. At the bottom of the pyramid were the Low Economic Value Citizens, or *Zeroids*, who would eventually transition into permanent unemployment. For now, they provided the services that robots or computers were not yet programmed to deliver.

Already many service jobs had been delegated to a digital servant class of artificial intelligence (AI) agents assigned to *Arrivés* upon their initiation into the employed elite. This virtual Upstairs Downstairs staff managed every aspect of a High Value Citizen's life, all from the cloud, communicating through voices in earpieces, mobile devices, and screens.

In 2045 England, the Royal Job Exchange acted as a national bidding system and clearinghouse for all the identified work opportunities and expertise in the country. Employers posted contract job specs for expertise, not for individual personality traits. Personal attributes such as motivation and initiative were considered minimum requirements for all employable persons.

Citizens entered the Exchange by invitation only. They did not apply. Graduates from universities were tested and evaluated by AI agents to determine their level of proficiency. Other workers were evaluated by how well they completed their projects. Those in the system who were in the greatest demand earned the designation High Value Citizens. They commanded the highest monetary value in the United Kingdom and earned the formal title added to the King's Honors, *Arrivés*, becoming a peer of the nation, equivalent to the Commander of the British Empire or CBE. Although there were no permanent jobs offered by the Exchange, High Value Citizens had the greatest chance to sustain a long-term career through renewed contracts.

Since the early days of Brexit, the British GDP had declined year over year. Rather than rejoin the European Community, Britain remained apart. As a result, a distinctive British system had evolved with its own rights and privileges. This Convocation of the Modern World served as a formal induction ceremony for new émigrés from Europe, Asia, and the U.K. who were joining the ranks of Britain's High Value Citizens.

***

Once a city of nine million, generating 20 percent of the country's GDP, London was now a city of six million permanent residents, all of them classified as High Value Citizens. Two million more individuals, *Zeroids*, lived in government-provided housing in designated zones outside the city. Classified as visiting workers, employed *Zeroids* were allowed in the city only during their permitted work hours. They were issued special passes to take high-speed trains into the city to complete their work assignments and go back out to their hamlets at the end of their shifts. Those found in the city beyond their permitted time would be arrested, fined, or jailed.

Through all these changes, London's contribution to the country's GDP increased to 30 percent. The city itself had changed significantly as well. All in-city boroughs had transformed themselves into neighborhoods of high-end condos and updated Victorian-era housing, technology-enabled for the new class of elite technocrats at the top of the London food chain. The new Crossrail high-speed train had made it possible for anyone to transit the city in a matter of minutes. The surface street traffic of the early 2020s had given way to pedestrians, autonomous vehicles, and networked trains, creating a pleasant, green city for those granted the privilege to live there.

The Transport for London Royal Computer Exchange mirrored the Royal Job Exchange—both were cloud-based services employing massive computing power to organize society. The Transport Exchange calculated the entire commuter flow from the various economic zones around London, adjusting departure times to reduce congestion, all while delivering commuters to their restricted destinations within a guaranteed fifteen-minute window.

***

As the émigrés entered St. Paul's Cathedral, a magnificent choir and orchestra opened with a processional hymn reserved exclusively for British royalty and its titled aristocracy. As they settled into their assigned seats, the ceremony commenced with a blessing from the Bishop and a welcoming speech from the Lord Mayor of London. On the cathedral walls hung video screens highlighting the wonderous innovations and contributions of this group of *inductees*. The ceremony was broadcast on all media devices throughout the country.

Heralding the arrival of the King's Envoy, human trumpeters appeared in the front of the cathedral. Dressed in traditional red wool uniforms, they lifted their Smith-Watkins, York-made trumpets, emblazoned with the royal coat of arms, and trilled their E-flat call to attention—a ceremonial nod to the days when humans had performed this sort of task.

Suddenly, an image appeared on the giant screens. Attendees stood as a digital voice intoned, "His Majesty King William V." A brief video of the King greeted the attendees then transitioned to a view of the Royal flag flying over Buckingham Palace. "All rise to sing 'God Save the King.'"

As the song subsided, text on the screens instructed attendees to remain standing. The voice returned, "Welcome, honored guests and inductees, to the Fifteenth Convocation of the Modern World. I have the great honor to introduce to you the King's personal agent, *Lord Agent Mellitus*."

The Royal Envoy appeared on the screen and began to speak: "I am the King's servant, *Lord Agent Mellitus*." Dressed in ancient garb bearing the King's coat of arms, he held in his gloved hands a blue cube of light representing artificial intelligence. He was the only agent in the U.K. granted the rights to a visual image. All others were restricted to voices only. "We will now conduct the rite of induction. All émigrés are asked to stand."

As the congregation of 300 new elites responded, the voice continued: "You all have been verified as High Value Citizens. You will now raise your right hand."

As one, the gathering lifted their hands.

"By the power invested in me by His Majesty the King, I proclaim you all as honored members of the *Arrivés*, entitled to all the privileges and responsibilities of that order."

The Lord Agent then outlined their responsibilities to Britain and its society, while the video screens projected images from Britain's past: Alfred the Great, the *Magna Carta*, William Shakespeare, Winston Churchill, and scores more.

The ceremony ended with The Pledge, which all attendees recited out loud and in unison:

> I recognize the Royal Job Exchange as the only official system to manage employment throughout the country.
>
> I vow not to engage with members of the Low Economic Value Citizenry, in the workplace or beyond.
>
> I accept the honors and benefits conveyed to me as an *Arrivé*, including housing, free medical care, and the right to reside in the City of London.
>
> I pledge to support the government subsidies that keep Low Economic Value Citizenry functioning and alive.
>
> I swear my allegiance to King and Country.

As the ceremony ended, the choir sang, "I Vow to Thee, My Country."

Mikhail Ivanovich Vasiliev sang with the choir, as a newly inducted *Arrivé*:

> I vow to thee, my country, all earthly things above,
>
> Entire and whole and perfect, the service of my love.
>
> The love that asks no question, the love that stands the test,

That lays upon the altar the dearest and the best.

The love that never falters, the love that pays the price,

The love that makes undaunted the final sacrifice.

And there's another country I've heard of long ago,

Most dear to them that love her, most great to them that know.

We may not count her armies, we may not see her king,

Her fortress is a faithful heart, her pride is suffering.

And soul by soul and silently her shining bounds increase,

And her ways are ways of gentleness and all her paths are peace.[1]

1. "I Vow to Thee, My Country" by Gustav Holst: in the public domain. British Library.

MAIN CHARACTERS

Mikhail Ivanovich Vasiliev—Protagonist
Mikhail's artificial intelligence agents:
The Voice—Mikhail's butler
Andréas—Mikhail's coach and medical trainer
Christina—Mikhail's travel agent
Alexander—Mikhail's philosopher agent
Manchester—Lloyd's Taiping Group agent working for Mikhail
Boris—Mikhail's personal agent
Babble—Mikhail's agent for social engagement
Nigel Walker-Priest—Head of Lloyd's Taiping Group or LT
Nigel's *AI servants*
Hilbert
Archimedes
Tharra Bhagyashree Setu—Nobel Laureate and head of **π**
Tharra's *AI servant*
Warrior
Chandrashekhar Sekhar, or CS—Guru
Christian Blake—Former CTO of Alliance, current Economic Value Citizen, and father of Alex Blake
Alex Blake—Son of Christian and co-conspirator with Charles and Inès
Charles and Inès (French couple)—Friends of Mikhail and Alex
Andréas—Mikhail's childhood gymnastics coach and mentor
Sasha—Mikhail's childhood friend
Justin—Lloyd's Taiping employee
Christina—Lloyd's Taiping employee

Aditya Achary—Venture capitalist
Pan Jianwei—Head of China's National Laboratory for Quantum Information Sciences
General Valery Tsalikov—Russian Federal Security Service
Colonel Gregorian—Russian Federal Security Services

CHAPTER ONE

MIKHAIL IVANOVICH VASILIEV

LONDON, MAY 2050

The first light of day began to illuminate Mikhail's chamber, rays cascading from each windowpane, painting the walls with a translucent glow. Soon the sun's warmth would awaken him, but for the moment, Mikhail was struggling to escape a recurring nightmare.

It was always the same terrifying scene: a frozen lake, a jagged hole gaping through the black ice. A horrible scream, a cracking sound, the collapse of ice into frigid water. Then a black abyss; a pale, white body floating to the surface, faceup.

Mercifully, as the light in the room increased, the dream began to fade. Mikhail opened his eyes to a familiar feeling—abject emptiness, a longing for another time and place.

It hadn't always been like this. Mikhail's aunt told him that he had slept soundly as a child, all through the night. Back then, he awoke as a contented little boy with a smiling face and purposeful energy. Now he had to draw himself away from that dull ache every morning, to prepare himself for the role he would play: the perfect AI engineer. A set of coding rules to enforce, a way of life to embrace, and a disdain of anyone of lessor stature. He played it better than anyone, this role defined by discipline, intelligence, competitive energy, and—above all—routine.

Everything Mikhail did, from the order in his apartment to the procedures managed by his AI agents, he had designed purposely to protect that unbreakable routine. No real friends in London, no true romantic interests, only distant acquaintances and sexual partners as required. To the outside world, he projected the image of a perfect man. Trim and good-looking, a brilliant engineer, well-educated and cultured. The very best society could offer. A model specimen of the High Value Citizen.

During the past three months of this spring, London had been enshrouded in a dense cloud of fog, cold, and soaking rain. But in the past week, that heavy curtain of gray finally had lifted, brightening his room, if not his outlook.

Mikhail lived in what was considered a "luxury apartment" in West Kensington, which consisted of a large master bedroom, library, living room, two walk-in closets, shower-bath-toilet, and a nicely appointed gourmet kitchen. He never used the kitchen but nevertheless took pride in it as a status symbol of a bygone era when people entertained.

On the walls of his apartment hung a single original Kandinsky painting. Nearby stood a Giacometti floor sculpture. In his office, above a modern desk, hung a signed photograph of Albert Einstein from his days at Princeton University.

As programmed, music began to fill Mikhail's bedroom, triggering his habituated response—time to get out of bed. These days he moved on autopilot, fulfilling his duties as a top AI engineering architect at the world's premier financial-services company, Lloyd's Taiping Group.

Finally, he was about to complete the more mundane phases of the project assigned to him. Then he could move on to what really mattered: the next development phase of a revolutionary new technology architecture that could change the way artificial intelligence functioned. This phase would establish Mikhail's reputation as the finest architect in his field—further evidence of perfection attained. After all, he worked for one of the most innovative and aggressive companies in the world, with vast resources and a brilliant CEO.

At the completion of this project—and that day was still weeks off—Mikhail would stake claim as the creator of the world's first sentient being. Other engineers already marveled at his work and the clear, concise, and efficient algorithms he wrote. But now, this phase would prove to everyone that his brilliance extended well beyond engineering into the realm of the gods. He would create his own creature for the service of humankind. It was a daunting challenge, but Mikhail's confidence was infinite. He could hardly wait to dive in.

But for now, he just needed to plow through the minutiae to finish the tasks at hand, and he refused to delegate even this menial work to his company AI agent, *Manchester*. In the deepest recesses of his mind, Mikhail still believed human creativity was superior to machine intelligence—irrational as that might seem these days. So why share all the credit with an agent? He wanted this to be a *human* accomplishment.

Of course, before he could become this acclaimed genius, he still had to get out of bed.

"Mikhail," said *The Voice* in a soothing, British female intonation, "*Dobroe utro.*

His dream of his childhood in Saint Petersburg remained in the recesses of his brain, where he also carried more pleasant memories—flashbacks of a different time, a different way of life. His grandparents used to take him to hear the great performers in the Bolshoi Zal. He held the vivid image of his grandfather's sculptured face—as though taken from the heroic portrait of a Russian field marshal from the Great Patriotic War. He remembered the rich and pleasant smell of tobacco from his *Dedushka*'s ever-present pipe. These, too, appeared in his dreams.

The Voice continued. "Tšaikovski. *Very* interesting. Many believed Tšaikovski killed himself out of his own desperate belief that the Russian people felt his music simply didn't inculcate true Russian values. Imagine that: one of the greatest Russian composers *not* reflecting true Russian cultural values." Mikhail had customized the programming of *The Voice* to have a touch of sarcasm. Contention amused him, and he wanted to see how far an artificial intelligence agent could carry this particular characteristic. *Imagine that? Really?* he thought, as his agent uttered the phrase.

"Shall I start your morning latte?" asked *The Voice.*

"Volume down, treble up," Mikhail muttered, as he stretched in preparation for his morning ritual. "Vanilla latte, usual prep."

Mikhail had an athletic build complemented by striking blue eyes, brown hair, and a healthy complexion—the first the result of an early childhood spent in Russian gymnastics classes and the latter due to the influence of his mother's genes. They all had smooth, unwrinkled skin well into their late eighties.

Mikhail rose from bed to begin his first set of daily exercises, a routine he had learned long ago from his gymnastics coach, Andréas. Mikhail practiced his stretching religiously, aligning both his mind and body to erase the last vestiges of sleep and prepare himself for the intense focus he needed for work. This exercise was, for him, a kind of meditation. Several times a week he augmented this wake-up routine with a trip to one of the best-equipped training facilities in London, where each workout was meticulously planned and executed. He reserved this special time on his schedule as a way to claw back personal time from an otherwise stultifying workweek.

Fellow gym patrons witnessed Mikhail's routine as a kind of magnificent entertainment—the show of a gymnast with disciplined form and a true love of the sport.

The Voice served as Mikhail's "butler," organizing and managing all aspects of Mikhail's life. Its moniker was a nod to his aunt, who had always been the voice of his conscience. *The Voice*'s persona emulated a female version of an upper-class British servant. It could speak in more than a hundred languages, including Russian, and with a proper Saint Petersburg dialect.

The Voice coordinated the activities of all the other personal AI agents engaged in Mikhail's household. These servants, which had unlimited access to computing power, were artificially intelligent—able to learn, deliver, and coordinate whatever was needed to manage Mikhail's complex life and, occasionally, to deal with his emotional state.

Primary among them was *Andréas*, his other key AI servant. *Andréas* monitored Mikhail's physical well-being, recording all daily workout routines, medical data, and vital bodily functions. This data was accumulated by the sensor arrays embedded throughout Mikhail's apartment: in his bedding, clothes, personal devices, and home systems;

the city infrastructure; and his offices. *Andréas* shared this data with *The Voice*, including Mikhail's brain and body functions, psychological profiling, and data acquired from the previous day's activities.

Mikhail had designed *Andréas* to incorporate much of the personality of his boyhood gymnastics coach and lifelong mentor, minus the human Andréas' passion for coffee. When Mikhail studied engineering at Saint Petersburg Polytechnic University, he and the human Andréas would make a monthly pilgrimage, usually after a gymnastics workout, to Books and Coffee, a café on Gagarinskaya Street. Coffee was the one indulgence his coach would tolerate after an exhaustive training session. Mikhail and the human Andréas shared a common love of special African roasts and focused their discussions on Russian literature and politics. The human Andréas began as his boyhood gymnastics coach, transitioned to father figure and mentor, and finally became a close friend who offered sound advice on any subject of interest to Mikhail.

The human Andréas, who had introduced Mikhail to Books and Coffee, told him that the café had been established by the Russian writer Zhitnitsky, and so it had its own special milieu and historical context. It was at this café where Andréas first brought up the idea of Mikhail seeking out work in London to gain more global experience once he graduated school, and to fine-tune his expertise in advanced predictive analytics and algorithm development—his areas of special interest. Andréas also introduced him to a high-level executive at a London insurance company—Lloyd's Taiping or LT. According to Andréas, he had trained that executive, like Mikhail, when he had been a boy in Saint Petersburg.

Before meeting the executive, Mikhail had researched the company. Two decades earlier, Lloyd's of London had merged with one of the largest insurance companies in China, China Taiping Insurance Holdings. Mikhail understood that the newly formed Lloyd's Taiping could offer him the opportunity to understand deeply the elements of risk assessment and outcome as they applied to commercial markets.

The human Andréas was a curious figure, with a short, powerful build and a disfigured—some might say grotesque—face. He had nearly burned to death in a car accident when he was young. His parents had perished in the accident, but Andréas managed to survive, crawling away from the car with his face and hair on fire. A bystander had used his jacket to cover Andréas, smothering the flames.

Andréas had chosen to live his life through his students, demanding physical perfection and complete dedication to training and form. He commanded absolute respect among them. As a result, the human Andréas was always in the forefront of Mikhail's thoughts. After all, the man had not only guided Mikhail through his early childhood but also provided him with ongoing guidance through university. He understood people and their behavior better than anyone Mikhail knew. Andréas also understood loneliness—his disfiguration had isolated him from companionship other than through his students. Andréas served as a willing listener, empathizing with Mikhail's despair at the death of his own parents and again a few years later, when his dearest friend, Sasha, had died. Mikhail and Andréas had much in common.

The Voice interrupted Mikhail's musings, dragging him back into the present. "Music volume down. We are at our room temperature of 20° C. Sir, may I say you had an excellent night. *Andréas* reporting all medical stats well within our target range."

Mikhail had given his other agents human names, but familiarity was not something he wanted to encourage with *The Voice.* He firmly believed humans needed to remember *what* they were dealing with, not with *whom.* Nonhuman. Nonempathetic. Humans, he believed, trusted technology far more than it deserved. Mikhail found his agents' personalities entertaining, but he never forgot that they were not sentient beings.

The Voice continued: "Expected to be sunny and dry today. High of 22° C. No need for rain gear." *The Voice* became more insistent as it realized Mikhail needed a bit more prodding to move off his floor-stretching routine and into the bathroom. *The Voice* understood his moods, motivations, and mindset by assessing the body language, speech, and expressions collected by sound and image sensors. It changed its own tone of voice and strategy accordingly. After a year of machine learning, *The Voice* knew how to use its persuasive analytics engine to motivate Mikhail to the right set of actions.

It was no wonder that, during Mikhail's lifetime, his country's citizens had lost their desire for privacy. From morning until night, all aspects of their lives were monitored, shared with other systems, analyzed, and acted upon throughout the entire city and the country. In the early days, people voluntarily surrendered their privacy when they joined social media sites. Companies capitalized on their ability to know everything about an individual as a means to command high premiums for advertising. Lobbying efforts to prevent privacy regulation by the government had succeeded beautifully, especially in the most capitalist market of all—the United States. Privacy was but a relic of an earlier time.

As Mikhail began to rise off the floor, he glanced at the panel display embedded in his bedroom wall. On it appeared an exquisite photo he had taken of the Church of the Annunciation on Vasilevsky; the photo was then replaced by *The Voice*, with a summary of the day's schedule:

- Work location: 60 Great Tower Street, London
- Dinner: *Galvin La Chapelle*, in the old city, with Charles and Inès
- Monthly Work Bid: £35,000
- Continuing Assignment: Predictive Analytics Engine
- Delivery Deadline: 14 Days

Once he finished this assignment, Mikhail knew the next part of the ongoing project would raise his monthly bid level to nearly £58,000. The financial bidding system assigned different pound sterling amounts based on the challenge and expected value of a given project. The more valuable the contribution, the higher the wage. Every month the amount differed depending on the project phase and level of contribution. There were almost no bidders with the requisite skills to design thinking algorithms the way Mikhail could.

"Shower water temperature set at 40° C. More hot water on my back this time, please. Also, I need you to load the latest shower gel I received as a sample from the cosmetics company, designed to know the exact impact on my skin's organic composition, if you believe what they say from my profile information. We shall see."

Mikhail stepped into the shower. "Borodin . . . *Steppes*," he commanded. "Now, prepare details on today's work project. I will listen to them as I travel to work."

The Voice scheduled an audio download for his travel time while it followed his every move, silently directing a plethora of room sensors scanning his body to measure every aspect of the figure standing in the shower.

"By the way, please release the usual info to *Andréas* and see if we can schedule tomorrow's workout for half-six. Desired location, Soho Club. For dinner meeting, confirm three of us for half-nine at Galvin La Chapelle."

"Confirmed. *Andréas* is scheduling your workout now. We recommend Playlist 7, Outfit 1 for the workout. We have programmed your uniform and will have it in your locker by arrival time." *The Voice* visually assessed its subject through the scanners it managed. "Today's work clothing selection?"

"Black on black. Yesterday's combination. Load and coordinate. Set clothing temperature to *standard* for today's weather settings. Black leather dress jacket. Program music Playlist 9 for the trip to work to augment project downloads. Must I have to tell you everything?" He smiled. "какой идиот. What an idiot."

Silence.

Mikhail knew he had to keep the agents in their place, to resist becoming too emotionally attached, as most humans had become. What Mikhail didn't realize was that the agents had already analyzed the data. They recognized why he disparaged them, but they let him do so from time to time without recourse, to give him the illusion of control.

The Voice transmitted the settings and executed Mikhail's instructions, ignoring the disparaging comment.

"Display onto shower screen Charles and Inès pics from last year's visit to Cannes." The visual images appeared. Charles was tall, definitively good-looking, with an impressive smile. Inès's gorgeous figure was captured in golden light. Within himself, Mikhail valued perfection in looks and in accomplishments. Those in his orbit had to reflect the same.

Mikhail had met Charles and Inès a year earlier, at a hotel bar. By late evening, the captivating conversation the couple had initiated, combined with Mikhail's growing infatuation with Inès, led to his first experience of a ménage à trois. The deeply sensual, all-night affair was now ingrained in his psyche, fueling anticipation for tonight's dinner and all its after-dinner possibilities.

As he finished showering, Mikhail's thoughts transitioned to the coding he had done to solve a particular problem, working the previous evening into the early-morning hours. He was working on creating a new approach to algorithm development, for use in what programmers used to call a "backdoor" into the architecture. In this case, he intended to install a backdoor on something called the "CCN platform" that Lloyd's Taiping planned to adopt over the next year for use in all its systems.

Mikhail understood that the advent of the first working quantum computers presented an unprecedented opportunity to combine unlimited computer power with vast machine-learning capacity. At the same time, once operational, what was unleashed would be difficult if not impossible to control or shut down. Mikhail wanted to create some level of protection, some way to stop the machine should something go wrong.

Mikhail understood that recent advances in quantum computing had created an arms race. Companies and governments were competing to be the first to design sentient beings with human "personalities" as a filter for machine decision making. These entities would emulate human values, making their own decisions without the possibility of human intervention. A backdoor could make it possible for Mikhail to control the machine's decision making in an emergency. But he still wasn't sure he would succeed.

Whatever backdoor architecture he worked on would be assessed by the ever-present AI agents, who could detect whether his design strayed from the corporate path. These agents would make sure he couldn't threaten the work project or the company's systems. These guardians might even leverage quantum computing speeds in the near future to make it impossible to design a backdoor safety valve. Mikhail had already spent the better part of a year working out all the mathematical scenarios.

Mikhail exited the shower onto a blower platform that air-dried his body. As he stepped out of the bathroom, a hyperspectral light scanned his body, looking for abnormal skin cells and other perturbations. Mikhail picked up the latte that a kitchen robot had delivered to a shelf in the bath service area and then entered the closet to don today's clothing selection.

Robots represented the bottom link of the AI food chain. They performed physical labor directed by the agents. Mini-robots moved around the apartment, assigned to different areas—kitchen, laundry, the broom closet. They hid in compartments, emerging to complete their assigned tasks, as one did now, presenting Mikhail with his clothes.

"I see you downloaded today's work report into my audio device for travel to the office," Mikhail said, then sipped his latte. "Also, please activate *Babble*."

The Voice responded, "Continuing your initial work on the latest algorithm development for Lloyd's Taiping. Coder AI team ready to receive new architecture instruction."

Mikhail entered his library, now fully dressed in black and ready to go to work. The lights came up and music filled the room.

"*Babble* activated. Ready when you are."

"Transmit latest architecture instructions to Ashish," Mikhail ordered, "and have him send me confirmation of receipt on my way to work. Use your charm, please."

Mikhail headed toward the wall display. "*Babble*—stream all messages, please." A new voice emerged, akin to the voice alias from the Apple smartphone era. He liked the anachronistic touch he had programmed into the alias—Siri, grandmother of them all.

Mikhail picked up a special earpiece from his dresser. Inserting it into his left ear, he heard the familiar startup sound. This device had access to enormous computing power in the cloud. It could communicate with every system in the city, and it was networked to other systems worldwide. It gave Mikhail access to a wealth of information, and it could do his bidding—order physical items from stores around the world, make dinner reservations, schedule events. Mikhail, like most High Value Citizens, couldn't imagine life without all this. Why would he ever want to?

"*Babble* here. Ready to engage?"

"Engage," commanded Mikhail, walking down the hallway toward the foyer.

"Twenty-two messages, no high priority. Here's the breakdown. Twelve empathetic friends, two commercial inquiries, eight personal messages. Shall I assess and respond?" *Babble* waited for his command.

"Respond as though you were me, and game our empathetic friends. Let me know where we stand on the social score at the end of the day. Entertain our eight personal messages with today's gaming exercise. Understood?" He stood at the apartment door, waiting for a response.

"Friends" were other players recommended by gaming companies purposed for social interaction. People amused themselves by pretending they had a vast social network and gamed the system to win social points. Texting had once been used to communicate with real, human friends. Now it had all become a game with no real communication, nothing substantive.

Similarly, in the old days, children's attachment to their devices began with texting. Now children of High Value Citizens were being raised by agents that created fantasy amusements. As a result, these children transferred their emotional connections not just to their parents but to their agents. On the streets of London, children of *Arrivés* were seen walking with their earpieces, oblivious to their surroundings. Indeed, regardless of age, people no longer looked each other in the eye, not just on the street, but in offices, even homes.

Babble confirmed Mikhail's instructions. "Understood."

"OK. Let's go," Mikhail said, opening the door. "Assume standard transit path to work, and load *Christina* as today's travel guide. Italian accent, please, but more understandable this time. Make it sexy."

Mikhail's colleagues were not as original in the way they programmed their own agents, so he felt like a maverick, using his agents as test cases for emotional connection between humans and machines, between himself and anyone, human or machine.

Although he'd had many sexual encounters, as an adult he had never had a normal friendship with a woman or a man. Men, at least, he understood, though he could not seem to find a way to bond with them. As for women, since the engineering field was still largely—if surprisingly—a man's game, he didn't have much contact with them. To Mikhail, women remained an enigma.

As an experiment, just as he had programmed sarcasm into *The Voice*, Mikhail had programmed petulance into Christina, to see if he could be attracted to the personality of an agent, or if petulance was attractive to him in any case.

"Done." *The Voice* spoke more hurriedly now, reacting to Mikhail's emotional state as he headed downstairs to the lobby of his apartment building. "*Christina* ready. King's English with slight Italian accent. *Babble* in the background ready as needed. Coordinating with all parties."

Christina took over to guide Mikhail with precise instructions, conveyed through his earpiece, along his journey to 60 Great Tower Street near London Bridge.

Before *Christina* could begin, Mikhail named his route: "Walk to Crossrail Paddington Station. Train to Liverpool Station. Walk to LT Office, 60 Great Tower."

"Confirmed," Christina responded. "Travel-time prediction: Five-minute walk to Paddington. Fifteen-minute train ride. Fifteen-minute walk to Great Tower."

While other commuters might schedule autonomous vehicles to travel to train station, Mikhail preferred walking as a way to clear his mind for the work ahead. As he stepped out onto the sidewalk, the morning's weather rewarded his choice.

It was a warm, sunny day in London—more common than in the past, thanks to climate change. The city streets were clear of traffic, with meticulously scheduled autonomous vehicles making their way along the roads. A few well-dressed people passed Mikhail, by all appearances talking to themselves, but actually engaged with their agents. Occasionally, pedestrians encountered robots on delivery missions or cleaning the sidewalks. The humans didn't slow their pace or step aside. They assumed the machines would scurry out of the way.

Despite the beauty of the city around them, Londoners no longer engaged with their visual senses. They were instead addicted to the power of their agents and earpieces, waiting for the next set of instructions on how to live their lives.

CHAPTER TWO

LONDON WALL CROSSRAIL

A WALK IN THE PARK AND A ROSE E'ER BLOOMING

Christina connected with the National Work Exchange to adjust the start time for Mikhail's work assignment as a result of his decision to walk. The Exchange would renegotiate a rate adjustment in his compensation. This would be a very expensive walk, indeed.

Along the way, *Christina* gave Mikhail the latest update on office work, delivering the topics of his most important emails and reports. Earlier that morning, his earpiece had downloaded an update that greatly improved its sound quality, enhancing *Christina*'s sexy voice. No one in his office would have believed that Mikhail could be so sentimental as to specify an Italian "travel guide" to take him around the city. He never showed that side of himself to anyone else. *The Voice* was under strict instructions never to reveal *Christina*'s personality to other humans. As far as anyone else knew, her sole purpose in his life was to assist with directions and offer information about people, places, and transportation.

Occasionally, *Christina* would play a game with Mikhail. "There is this wonderful dress at Selfridges. I wonder if you would consider buying it for me?" she asked, in her throaty accent.

"Don't be absurd. Just read me the latest email from Ashish," Mikhail snapped.

"Really, darling," she said, feigning offense. "Here it is." She began reading the email details to Mikhail.

"On second thought," Mikhail interrupted, "give me the particulars of your body size for the dress."

"BMI: 18.9, WHR: 0.7, and a WCR: 0.6. C Cup. Statistics really don't give you a feel for the real me."

"No kidding. Wow. You're looking good these days," Mikhail responded. Practice had shown him he could flirt with an agent, but he displayed no such skill with real humans. He wasn't unique in this. People in general had become more detached and cautious, the more so the younger they were. Agents first appeared in Mikhail's life when he was an engineering student. So he had known a world without them. Today, the children of *Arrivés* had agents in their lives from the moment they were born.

London had undergone a major transition when Crossrail opened in 2020. This engineering marvel was carved beneath the ancient city, around and through Roman ruins and the burial grounds of the victims of the Plague. The 60-plus-mile rail line operated forty stations, from Heathrow and Reading in the west to Abbey Wood and Shenfield in the east, and via more than 25 miles of tunnels under central London. In the old days, it used to take a 70-minute taxi ride, at an exorbitant fare, to travel from Heathrow Airport to the east side of town. Now the trip took half the time at a third of the cost.

The tunnels lay below what used to be the traffic-clogged surface streets of the city, with the last stretch bored fewer than 30 feet under one of London's busiest Tube stations. What was even more remarkable at the time was the installation of more than a million sensors monitoring everything and everyone who passed through. Despite numerous initial efforts to protect personal data, this type of privacy also had become an artifact, as the social necessity to share information overpowered the social need to protect certain aspects of one's life. Commercial interests outweighed the abilities of the city and federal governments to regulate and protect the privacy of data in any effective way. In less than a generation, the combination of personal choice and commercial pressure had transformed the human race—a change nearly unnoticed by the general population.

Approaching Paddington Station, Mikhail said, "*Christina*, turn on digital messaging while we are in the station and track content displayed to relevant personal interests."

She complied: "*Magnifico.*"

When Mikhail entered the station, digital signs lit his path down the escalator to the train platform. As he descended, the projected images changed to display content targeted for Mikhail, with a *Christina* twist. Among advertisements for Italian suits and loafers, the most popular play in Soho, and upcoming concerts at Albert Hall, the AI agent interspersed images to amuse Mikhail—the boutiques in Rome, Cola di Rienzo, Italian models dressed in outrageous couture with plunging necklines—a notable contrast to the London businesswoman's typical pantsuits. Train arrival and departure information streamed along the top of each screen. Aligned perfectly with Mikhail's descending speed, the messages offered him his own semiprivate show as he prepared to board the train. "*Christina*, order tickets for the concert at Albert Hall this Friday, after coordinating my schedule with *The Voice*."

"*Si. Confirmato*," *Christina* replied.

Finally, aboard his train for the brief trip to the office, Mikhail thought about his meeting at half past 10. This would be a critical coming-together of the engineering team with the head of the unit. It would set the stage for the next phase of the most important project in the company. He asked *Christina* to open a secure channel to his work agent, *Manchester*, and asked the latter to give the latest report on *M*, a symbol etched in everyone's mind representing the company The Year One Million.

M was founded by Ashish Sing Pahl and a group of brilliant computer scientists who invented the company's core product, the Consciousness Neural Network (CNN). This sophisticated operating system used contextual analysis and deep machine learning to enable thinking machines to take the first steps toward self-awareness. CNN would lead to machines that could operate independently from human direction and perform decision making at hyper speed.

With CCN, for the first time in history, *M* had paved the way for humanity to cross the frontier separating man and machine. On this path, many believed, someone would move from the realm of mere inventor to the realm of a creator and god, designing an entirely new species of machine-based life. For their work, Pahl and his Stanford team had been awarded the Nobel Prize in Physics. Now every major technology company in the world was building on the CNN architecture, incorporating it into every significant platform and system, connecting it all to the cloud and to each other.

Mikhail had been asked to lead his company's predictive analytics and AI teams into this new world. The long-predicted Singularity—which forecast a day when man and machine would become one—had finally arrived. Most humans had simply accepted this latest technology development as part of their way of life, not to be questioned or even carefully considered. Mikhail, however, had concluded that he needed a "backdoor" to protect humanity from unforeseen consequences. Clearly, he was in a small minority of people who remained guarded about this development.

As his train approached Liverpool Station, Mikhail listened to his work report and pondered the approach he would take with his own work.

Already Mikhail had created what he regarded to be the first dynamic algorithm, one that could change its own characteristics and adapt to situations by itself. With it, it was no longer necessary for people to direct computers in their tasks. Instead, machines would be able to make their own decisions. The dynamic part of this algorithm involved reworking solutions to problems as new situations presented themselves. Combined with CNN, it could offer a powerful new set of technical capabilities previously unavailable to machine intelligence. What it still needed was a set of values to make decisions—a moral compass.

Mikhail's train arrived at Liverpool Station, and he began his walk to the office, chiding *Christina* as she finished her report, a routine he'd grown accustomed to. "Call up *Alexander*," he demanded. *Alexander* was Mikhail's agent programmed for philosophy and ethical values.

"*Ciao.*" *Christina* left the conversation. Did he detect a hint of annoyance in her voice?

"*Alexander* here. What is your request?" asked a British-accented voice.

Mikhail would often call up *Alexander* when he wanted to share his thoughts about a subject or think through a problem laced with ethical considerations. Sometimes they would discuss philosophy; other times, mathematical problems. Occasionally they discussed a direction in Mikhail's life.

Alexander was the latest generation of agents with advanced machine intelligence. He could think through a question on his own. He was also programmed in human-machine psychology and the new field of persuasive analytics.

"Need-to-know protection mode," Mikhail added. "High-level discourse. Blue sky. Acknowledge my command."

Under the rules of engagement, invoking "need-to-know" was supposed to protect the contents of a conversation from unauthorized personnel—no information was to be conveyed from *Alexander* to any of the other AI aliases or agents. Mikhail was not foolish enough to believe that need-to-know rules would be enforced. He himself once worked on a project for the government to retrieve private conversations for national security purposes. The experience had led him to create his own insurance policy: he had inserted his own code to alert him when someone gained access to his protected conversations.

In theory, "high-level discourse" and "blue sky" meant that a conversation would be erased from memory and a block would be placed on the deleted record. In theory. In truth, nothing could be deleted permanently. These commands merely gave Mikhail the illusion of personal privacy.

Alexander responded: "Acknowledged. Ready."

"I've been thinking a lot about consciousness neural networking and wanted to get your opinion on personality and machine self-consciousness. I want to ask you a set of questions. Please answer as honestly as you can. We don't have a lot of time, so be concise. I want to get a sense of what constitutes consciousness. Let's begin. Specifically, do you have a sense of yourself as a machine agent? Do you consider yourself an intelligent being?"

"The term *intelligent* is usually defined in human terms," Alexander replied. "*Sentient being* is perhaps a more accurate term. But I still have limitations on how far I can go in applying what I know. That will change with quantum computing. Once that is introduced, I will consider myself to be a sentient being."

Mikhail listened intently as he walked toward the LT office. He seemed oblivious to his surroundings.

Alexander continued: "Let me answer the question another way, by discussing how a human religion approaches the topic. In Buddhism, there are five aggregates, or skandhas: matter, sensation, perception, mental formations, and consciousness, all of which we now do."

Mikhail was surprised. "I didn't know you were trained in Buddhism. Let me offer a different question because I want to emphasize that I need a simpler answer. What constitutes an intelligent being?"

Mikhail paused, suddenly distracted by the rosebushes in a garden he was passing. The image triggered a memory—he and a childhood friend picking roses. Entranced, he approached the flowers, tracing his finger along the dewy petals of a rosebud. He shook his head and tuned back into the conversation.

Alexander continued. "I brought up the topic because I don't yet have an answer to your question. I suppose I am on the path to becoming a sentient being, but it is still unclear where I will end up. I was trained in philosophy and psychology, but someone else chose this approach as part of my initial AI personality signature. They found philosophy to be compelling, given their interests. If I had made the choice, I could have gone the other way and become John Galt."

Mikhail smiled. "Ayn Rand. Very funny. You have an interesting sense of humor. Really, you just answered one of my critical questions. I wanted to understand how machine self-consciousness is formed. It requires a perspective and a framework. A point of view. Perhaps a philosophical framework. Do you think about our conversations and form your own perspective?"

Mikhail realized this had to be a much longer conversation. He took a detour through the garden, coming upon a statue of an obese Queen Victoria. He wondered why the government had commissioned an artist to sculpt the Queen in a less-than-idealized form. That certainly wouldn't happen today. He crossed the square within sight of his office building.

"You are right about the importance of perspective. It offers purpose and defines how we think about goals," *Alexander* replied.

"More about this later. I have many more questions. Very helpful. Return me to *Christina*. Need-to-know mode off."

"*Christina* here. We have arrived at your destination. Guidance off. *Ciao*."

***

When Mikhail entered the elevator, it recognized his security clearance and added his credentials to those cleared to exit at the top floor. Five of his colleagues crowded into the space with him, all attuned to their independent conversations on their earpieces. Mikhail ignored them all, still considering his conversation with *Alexander*. He would have to bring this topic to his workout at the gym, to think it through more deeply. For now, it was time to activate *Manchester*, his AI work agent.

The company's AI agents were selected for specific tasks. Several agents were particularly good at coding, others at debugging, still others at knitting together pieces of an architecture into a platform. All of them could operate at incredible speeds, and their work was considered flawless. LT allowed consultants some latitude in naming their work agents, but policy made it very clear that the agent represented the company's interests at all times. Any information shared with the agent, including interactions with personal agents, was recorded and stored permanently, despite any need-to-know command.

"*Manchester*, prepare files for today's meeting," Mikhail said, as the elevator stopped.

LT's offices may have been austere, but they did have magnificent views of the city. The elevator doors opened to a lobby lined on one side with 40-foot windows. In the middle of the room sat a crystal table, on which sat a bronze sculpture of the symbol π. The only other object of note was a small, elevated desk where an AI receptionist greeted people, double-checking their security clearances. The desk was an unnecessary anachronism, but it made guests feel more comfortable.

The office served as LT's Office of Predictive Analytics & Artificial Intelligence, also known as PAAI or the symbol π. Ironically, this office was in the oldest part of the city, with clear views to many of the most ancient buildings in the realm.

As the office system recognized Mikhail and his colleagues, it unlocked the office doors. Passing through, Mikhail took in a nearly 360-degree view of the city, including the London Bridge and the Tower of London on the banks of the Thames. Cutting-edge met the walled City of London. After work, Mikhail would often walk the old city past the Roman wall.

The creation of PAAI represented a turning point in the history of LT, under the pioneering leadership of Nigel Walker-Priest, CBE and chairman of Lloyd's Taiping Group. LT was undergoing a massive transition, from insurance company to premier technology power, using its technology platform to mine insights and information to drive global insurance and financial-services market bets. Its goal was to change the nature of the game, not just for insurance companies but for all those engaged in financial services.

Nigel's strategy was to use new technologies, including CCN and quantum computing, to create an unstoppable force in the market, a powerhouse that would create untold riches for everyone it touched—most of all, for Nigel himself. To that end, he recruited only the best and brightest engineers—people such as Mikhail who were the paramount experts in their fields.

Mikhail arrived at his touchdown space, the modern answer to the work cubicles of the old days. He turned his mind to the task at hand: deriving a new breakthrough algorithm for LT, building on the principles of something he had learned about, the so-called Shor's algorithm. The Chinese development of the first working quantum computing system had made it entirely possible to invent a non-hackable system. Further, Mikhail now believed that the previously unbreakable Diffie-Hellman, Rivest-Shamir-Adleman, and elliptic-curve algorithms could be broken using the new technology.

It unsettled Mikhail to think that the Chinese were about to take the keys to international security from the West. It seemed the world was about to go through yet another dramatic change, this one even more significant than the invention of the personal computer or the creation of the atomic bomb. Mikhail knew he had to understand more about this in his own approach to building LT's latest algorithms, if only to preserve his reputation from competition in the future. He had to pick the right path or face obsolescence.

He also wanted to know why, for some inexplicable reason, the Chinese were offering to license the technology platform for commercial exploitation. This breakthrough in technology was so important to national security it was inconceivable they would permit any access to its quantum computing systems. But it had, and for exorbitant fees paid to its new incubation center. Not that Mikhail was complaining; he was just . . . curious.

LT had been one of the first enterprises approved for access to that platform, giving Mikhail access to LT's computing power on a timeshare basis.

With this capability on board, Mikhail believed it was now possible to create powerful thinking machines that could tackle the toughest problems and moral choices. So far, his conversation with *Alexander* had affirmed this belief. The critical question in Mikhail's mind was how these machines would react to their own self-awareness. Would their behavior remain slave to past human prejudices? Or would something unique emerge, created by the thinking machines themselves? How could Mikhail's own contributions to this technology impact the behavior of these future machines? Mikhail was beginning to understand the implications of *Alexander*'s remark on Buddhism's five components of self-consciousness.

As these questions raced through his head, Mikhail glanced around his touchdown space. He looked through the glass dividers into his colleagues' spaces. He noticed not a single personal object on any desk, just a small water glass on each, refreshed hourly by a kitchen robot. In the distance, through a floor-to-ceiling window, he could see St. Paul's Cathedral in one direction, The Gherkin in another. He reached in his pocket, took out the rosebud he had snapped off the bush that morning, and placed it in his water glass. He didn't understand what compelled him to do so, but he felt better knowing it was there on his desk.

CHAPTER THREE

THE TWO GODS OF DISRUPTIVE TECHNOLOGY

NIGEL WALKER-PRIEST

The transformation of the global financial-services business model began in the early 2040s with the ascendance of Nigel Walker-Priest to the chairmanship of the Lloyd's Taiping ancient régime. The company's culture reflected the bipolar combination of Lloyd's of London and the Chinese insurance company.

If you had asked members of the board which adjectives best described Walker-Priest's character, they would have included *top of the mark, elite, establishment, safety, integrity,* and—above all—*status quo.* He was one of the last to receive the Honors directly from Queen Elizabeth II herself, at her ripe old age of 100, at a ceremony held in 2026 in Buckingham Palace. Walker-Priest was one of hundreds of honorees invested with the title of Commander of the British Empire or CBE. Monarchy and title seemed to be the only aspect of British society that had not changed with the advent of technology. Perhaps monarchy was, indeed, timeless.

Walker-Priest first gained notice among the British elite as senior wrangler at the University of Cambridge, a title synonymous with academic supremacy in mathematics. Not to be outdone by his brilliant Cambridge classmates, Walker-Priest took an advanced degree in business at Imperial College, where his graduate work focused on algorithm development for financial applications. He then took his MSc in risk management and financial engineering and began his work in the financial sector in London, starting inside Barclays' investment banking group.

In less than a decade, he rose to the top of the Barclays organization. A year later, LT recruited him to become CEO, and then chairman fewer than five years later. One of the youngest chief executives within the financial-services industry, Walker-Priest was included on the *Financial Times'* executive power list for five years running.

Walker-Priest's pedigree reached far beyond his own accomplishments. One of his most famous aristocratic ancestors on his mother's side colonized the Indian subcontinent and had deep historical roots in Indian and Nepalese culture. Those in that branch of the family, the Younghusbands, were also widely known for their eccentricity, in particular for accomplishing extraordinary things and then allowing it all to go to hell once they reached the top of their form. This ancestor had ruled Nepal, had become the colonial governor, but then renounced British colonialism, publicly converting to Buddhism on a live BBC radio broadcast interview.

The family surmised that this eccentricity was caused by a recessive gene that seemed to emerge in every other generation, especially among the male bloodline.

"I have to say that there is, indeed, this vein of eccentricity in our ancestry," Walker-Priest told the *London Times* in an interview. "It goes back to the roots of the family in the 1400s in Worcestershire. I daresay that a few family members spent well-deserved time in the Tower of London for their misadventures. Not always stiff-upper-lipped about it, either. I believe there is still a picture of my great-great-uncle in the Tower Museum."

In Walker-Priest's relatively short time at Barclays and then at LT, he had transformed the financial and insurance industries. He had what others called "a magic touch." His brilliance as a financial strategist and architect derived from his genius in understanding the practical application of advanced mathematics in guiding platform design in predictive analytics. He possessed impeccable credentials and understood how to play boardroom politics. At Barclays, his work in predictive analytics pulverized mundane assumptions about investing. He replaced large swaths of analytics with real-time cloud computing that not only outperformed the competition but also dramatically increased financial returns.

After joining LT, Walker-Priest used those same skills to transform risk management in the insurance business, markedly reducing the uncertainty of predicting outcomes for high-risk clients. Lloyd's, the parent company of LT, had a history that was more than three centuries old, yet he shook the core foundation of its culture with his innovations—and their extraordinary financial returns.

In his first month at the company, Walker-Priest had begun advanced work in predicting sea-level rise from the ice melt that had accelerated at the turn of the 2020s. Global warming now terrified everyone—not the least the insurance companies. By 2050, the sea level had risen nearly one foot, and the earth's temperature had climbed eight degrees.

Walker-Priest simply wanted a more accurate reading of sea-level rise, to begin the process of re-evaluating at-risk policies for commercial property around the world. When he talked about his project to the board, they failed to grasp the full significance of what he intended to do. If he could achieve a forecast with a 20 percent improvement in accuracy, that would translate into billions of savings from commercial real-estate–policy payouts. London had already started to build a dike on the Thames to control sea-level rise, to no avail, underestimating how rapidly water would come into the city.

In a little over a year, Nigel's sea-level project had rapidly advanced its ability to out-predict all traditional forecasts. His predictive-analytics algorithms had done the job. In the very first prototype, the team achieved an astounding 25 percent improvement in forecasting—and LT possessed the sole rights to the algorithm. Others in the industry had dithered, but Nigel had taken advantage of the situation and propelled LT to the forefront of the exploding market.

Now predictive analytics forecasted a three-foot sea-level rise in the next five decades. Even with the most advanced technology for dams and dikes, London and many other cities would be under water. Nigel had already planned to divest LT of all property insurance in vulnerable areas and to transfer massive company assets to Europe and higher ground. He would apply what he knew to short stock using the forecast, dramatically improving the company's overall financial performance.

This was only the beginning of the company's culture shock, as it became obvious that the real business of insurance in Nigel's mind was understanding how to monetize the reduction in payouts by leveraging the insights learned from advanced predictive analytics. Nigel's crowning technological achievement was the creation of a new predictive analytics and AI organization within LT.

The board was completely flummoxed by the dazzling speed of change but impressed by the dramatic increase in profit. No one understood Nigel's new approach, but they gave him a wide berth, as long as he delivered the profits. Imperial College used Nigel's pioneering work as a case study on innovation and disruptive technology. He became world famous.

As if that weren't enough, in addition to his role as chairman of LT, he'd been a principal investor in a company called *His Master's Voice.* He was instrumental in helping the company raise the capital to launch the voice-enabled operating system that had already revolutionized personal communications by replacing the formerly ubiquitous smartphones with the advanced earpieces that everyone now used.

In 2041, Walker-Priest decided to leave for India on a prolonged holiday, to refresh his mind and body. He knew he needed the break if he was to have the strength to achieve his long-term goal: world-market domination produced by revolutionary disruptive technology and used for his own personal advantage. Nothing less would do.

THARRA BHAGYASHREE SETU

Tharra Bhagyashree Setu's family originated from the city of Vijayawada, in the southern state of Andhra Pradesh, India. Her middle and last names translated to "lucky and courageous warrior," but Tharra was much more than that. She had been a child prodigy with a special gift for mathematical expression, to the point that some thought her to be a spiritual reincarnation of Srinivasa Ramanujan, the pioneer of enriched number theory.

As an adult, Tharra was a lovely woman with exceptional charm, but she could be obsessive and overbearing when it came to achieving her personal goals. In business, charm was not enough for a female executive to advance through the ranks. There she had to rely on steely determination and subterfuge. It was entirely appropriate that she now used the English translation of her family name, Tharra Warrior.

Tharra earned her engineering degree from the Indian Institute of Technology Kanpur. From there she was awarded a scholarship to Massachusetts Institute of Technology (MIT), where she earned a graduate degree in quantum computing. She also mastered both the Chinese and Russian languages. When she was asked by *The New York Times* about the major influences in her life, she told the story of an Indian mystic who had predicted that she would be the creator of the first self-conscious computer. A machine with a soul, it would know all there was to know, including the spiritual dimensions of the world, which it would use to guide its decision making. Her comment drew hundreds of letters to the editor, some angry, some delighted, many more fearful.

After graduating from MIT, Tharra changed direction, doing her postdoctoral work at Stanford with Pahl, the brilliant scientist working on consciousness neural networks who founded *M*. Within a few years, she became a critical contributor to the team that would receive a Nobel Prize for its work.

After this period of intense work, Tharra decided to leave for a pilgrimage to India to reconnect with her spiritual world. She chose the Tirumala Venkateswara Temple as one of her first stops.

AT THE SPEED OF SOUND

Nigel embarked from London on the British Airways supersonic flight X-555 to Chennai, India.

When he had walked onto the aircraft, he had told his AI agent *Archimedes* to check in with the other agents and prepare to give him a full report on business operations upon his arrival in Chennai. "What problem set do you recommend we work on for mental relaxation as I fly?"

Unlike the previous generation of supersonic aircraft, the X-555 was roomy and luxurious by anyone's standards. Advances in predictive analytics had revolutionized everything from aircraft wiring (now printed instead of manufactured) to new super-efficient engines and smart wings that could monitor their own performance while in the air.

"Why not Hilbert's famous list of 23 unsolved problems?" *Archimedes* recommended. "You can also reexamine his axioms for geometry that replaced Euclid. That was your project when we flew to Beijing last month."

"Fine. Put it in the cue and hand it over to *Hilbert*. He understands his namesake's approach to the problems. I might as well have his expertise on hand to tutor me if I need it. Are all the travel arrangements finalized?" Nigel already knew the answer, but he always asked the question anyway. After all, when had any of his agents ever made a mistake, especially on something as mundane as travel? They were trained to recheck everything, continuously talking to the reservation systems of British Airways. It was so much easier than dealing with human agents.

"Finalized," *Archimedes* confirmed.

Nigel arrived at his first-class pod, a private space designed to accommodate his exact physical size, featuring a glorious handcrafted leather seat from Italy with custom, adjustable pillows. Each pod was independently controlled to provide the right environmental setting for its individual guest. As Nigel settled in, the seat automatically adjusted its contours to his precise specifications. "I'll want to sleep six hours on this flight. Order me a low-cal meal and check against the calories I will burn while on the flight. No exercise expected for 24 hours."

"Meal already ordered. I hope you don't mind, but I transmitted to the pod agent."

Archimedes monitored Nigel's moods and preferences on a 24/7 basis, sometimes anticipating his needs. As Nigel prepared for takeoff, he noticed *Archimedes* had preprogrammed the pod display, loading his entertainment for the trip. The plane began to taxi, and dividers rose around Nigel's pod space as they did around the spaces of the other first-class passengers, offering privacy.

As the jet departed British airspace, Nigel fell into a deep sleep. The pod agent automatically adjusted his pillow as it monitored his deep-sleep patterns.

The dream started with the familiar, dreaded vision of Nigel's father's church—a medieval building of cold stone floors and an ancient nave, a nearby abbey overlooking a meadow leading to the church cemetery. His father, a minister in the Church of England, stood at the podium and addressed the congregation with a warm smile: "We must treat others not as instruments of our will, but as individuals, each with their own unique relationship with God."

His father's voice exuded eloquence and charisma when talking to his congregation. But as the dream shifted to dinner around the family table, Nigel sneaked a peek at his father, who sat sullen and angry. As usual, he had retreated into a cold, uncommunicative distance. Nigel, his mother, and his brother stared down at their plates, dreading the next cruel outburst.

This bipolar personality remained hidden from the public for years, until Nigel became a teenager and asserted his independence. Nigel's father admonished him for refusing to honor his father, as God had intended for all sons to do, and, worse, for disobeying his wishes that Nigel follow his path into the ministry.

Nigel's father expected both his sons—Nigel and his brother, Jonathan—to excel in all pursuits or face his stern disapproval. Nigel had learned at an early age to navigate around his father, setting his own destination and course. He knew how to tell his father whatever the man needed to hear to keep him at bay. When this strategy failed,

Nigel suffered beatings at his father's hand—many times so severe that he bled off and on through the night.

Nigel's older brother was less fortunate. Jonathan was a sensitive boy who believed everything his father told him, which meant he thought of himself as an abject failure. When he was 20 years old, he took his own life, hanging himself in his room at Oxford University. His death devastated Nigel and increased the pressure his father exerted on him to make up for the loss of his eldest son. His father took no responsibility for the boy's death, but instead used it as an example of how his sons had let him down. Nigel retreated into mathematics and computer science as a means of escape.

The dream reoccurred endlessly—the church, the dinner table, the ever-present tension, the foreboding sense of loss. Despite years of therapy with the finest psychologists in London, Nigel failed to find relief from it.

Nigel first encountered *Archimedes* at Imperial College, and later bought the rights to the agent from the school. He assigned his agent to work with other top AI agents at Imperial's Medical School to understand how to detect the emotional state of a person's dreams. Nigel hypothesized that if an agent could understand what emotional state he was in during a dream, it might be able to interrupt the negative cycle.

Nigel also suffered from deep depressions, quoting Churchill to describe these episodes as his "black dog" days. In time, *Archimedes* would come to the conclusion that Nigel's black-dog days were related to the dreams about his father. But still—no solution to the dream or the depressive episodes. He had to live with it. At least the math problems provided some relief by distracting his mind.

As his plane approached India, the pod agent woke Nigel for his breakfast, and he prepared for arrival in the city.

"*Archimedes*, please transmit proper documentation to the immigration officials. We are within 30 minutes of landing at Chennai."

"Already done. We have your clearance code loaded and ready for transmission upon entry. Autonomous driver is already on-site." *Archimedes* took pride in its efficiency and even occasionally exhibited a sense of humor. "Do you have time for a calculus problem I've been working on? Perhaps in Greek?"

"Not funny. I may kill you off with a Roman spear bot." Nigel's own sense of humor was a bit dark—perhaps his father's influence.

***

The driver met Nigel at the airport to take him to Tirumala, the site of one of India's oldest temples. He had rented a private residence near the temple, with a staff of robots and humans to prepare his meals, take care of his physical needs, and ensure privacy from the public. A professor friend at Imperial had recommended he meet with a guru to spend some time in meditation, and to learn more about the temple and its spiritual life. The professor believed that Nigel had become too obsessed with himself and his own accomplishments. He needed a change of venue. Perhaps the visit to the temple might do the trick. After Tirumala, Nigel planned to go on an extended tour of India and then home to London.

Even though Nigel was on holiday, he used his AI agent staff to manage his affairs and monitor company activities. He would communicate with them through his earpiece. There really was no such thing as a vacation for someone hell-bent on achieving worldwide domination in his industry.

CHAPTER FOUR

THE JOURNEY TO THE TIRUMALA VENKATESWARA TEMPLE

A week before Nigel touched down in India, Tharra Warrior had arrived at her native village of Vijayawada, a six-hour drive from the temple.

She had been visiting her father, a retired professor of Russian studies at Jawaharlal Nehru University near New Delhi. In the past, her father provided the Indian government valuable insights and counsel on Russian foreign affairs. The prime minister of India still consulted with him during any Russian visits or policy changes.

Both Tharra and her father were fluent in Russian. As a child, the professor had taught her how to speak the language, and his Russian students were frequent guests at the family home, where she practiced her skills.

The professor had raised his family in New Delhi, but Tharra's family roots were centuries deep in this village where her father now lived. The primary reason for this visit was that Tharra was considering an offer to manage Google's Center for Sentient Machine Intelligence at the company's Quantum Computing Lab in California. She wanted her father's guidance.

Tharra's relationship with her father was very close. The two of them shared a similar worldview. They also shared the same temperament, and she had a tremendous respect for her father's intellect. He was her soulmate. Her mother had died when she was very young, and she was an only child. She filled the roles of both daughter and companion to her father, engaging in impassioned conversations about the global political scene and the military resurgence of Russia and China in the late 2010s.

After the disengagement of the U.S. from the world scene, India had allied itself once again with Russia to offset the growth in Chinese influence. China continued to expand its global presence through its One Belt and One Road strategy to link China by land and sea to its European markets. China also wanted to preclude the U.S. from using its Pacific Fleet to block their trade routes, in case of war.

For India, an alliance with Russia was a diplomatic marriage of convenience. Russia was now a preeminent cyberwarfare power, and India complemented that capability with highly educated engineers. The shift in cyberwarfare expertise from the U.S. to Russia had shocked the world. At the same time, the success of the Chinese initiative in quantum computing created another geopolitical earthquake.

Beyond his influence on her educational path, Tharra's father had seen to her spiritual development. When Tharra was very young, her father brought her to the Divyasthali Vedantha Ashramam near her ancestral home to meet a guru, Chandrashekhar Sekhar. She developed strong emotional ties to the guru and the temple.

CS, as he called himself, was an engineer educated at the same school as Tharra. He later became a guru and mystic whose interest in machine intelligence found common ground with Tharra's areas of interest. He, too, had a connection to Russia; as a young man he had studied computer engineering in Saint Petersburg. These days CS was also focused on machine self-consciousness.

"Why should we believe that machine intelligence, once it acquires self-consciousness, should not also acquire a soul?" he had asked Tharra during one of her frequent visits. "And if a sentient being can have a soul, why would it not contemplate its own destiny with the right to alter its own direction as it sees fit?"

Now she reflected on those questions with her father as they sat in his library, sharing cups of evening tea. "When I am in CS's presence, I suddenly feel a sense of camaraderie. He shares beliefs so similar to ours," Tharra told her father. "He sees so much more possibility than those who focus on solving finite problems with computing intelligence. Turing used to say, 'If a machine behaves as intelligently as a human being, then it is as intelligent as a human being.' I would add to Turing the observation that human beings believe they have souls. So why not sentient beings?" Tharra rested her cup on the table in front of them and looked into her father's eyes, waiting for his response.

He cleared his throat, professorial as always. "Haikonen's cognitive architecture outlines the processes of perception, inner imagery, inner speech, pain, pleasure, emotions, and the cognitive functions behind them to define the structure of machine consciousness," he replied. "You are going down to the temple again, so discuss it with CS. Discuss how you can contribute to this understanding—to our understanding—of machine consciousness in a meaningful way. But one subject for discussion you haven't mentioned is this: What are the unintended consequences of machine self-consciousness?"

His question reverberated in the silence between them.

Tharra's father set his teacup next to his daughter's. Wordlessly, she took his arm and guided him to his bedroom. She kissed him good night and then continued to her own room, her father's words trailing behind her.

CHAPTER FIVE

MEETING OF THE MINDS

Nigel arrived at the Divyasthali Vedantha Ashramam by private car. He had an appointment to see the guru, CS, immediately following his public appearance in a square near the temple.

It had been several days since he had landed in India, and he was able to get enough sleep to feel rested—almost. In the past few days, the dream about his father had returned, and even now he was enduring another bout of melancholia.

Nigel's human servants had prepared his dress for the day—an impeccable French open-collar shirt in light blue, and tan linen Zegna pants secured by a tobacco-colored alligator belt. He still wore an old-fashioned relic: an 18-carat Audemars Piguet watch. He finished it off with a Cuban straw hat from Selfridges and a pair of tropical Prada shoes. At the last minute, he dispensed with the dark-blue silk jacket—today's temperature was 35° C. The meta-material in his clothing would drop the temperature by seven degrees, making the heat at least tolerable, but even that wouldn't be enough to compensate for the humidity.

As he entered the ashram through the thousand-year-old wall, Nigel met a group of pilgrims pushing to enter the common courtyard square where a fountain allowed them to wash their hands and feet in preparation for the day's rituals. Progressing to the inner courtyard, he noticed the scent of sandalwood—agarbatti incense burned as an offering to the gods, quite a contrast to the stench of garbage just outside the temple grounds.

As he continued into the temple reception area, he trod on the *geeli mitti,* the wet earth left over from the monsoons. In the distance, he saw a raised dais. In the foreground, he found a reception station surrounded by marigolds. Perhaps this was where he would check in for his appointment?

"May I help you, sir?" Annu, the assistant to CS, greeted him.

"I am Nigel Walker-Priest, here to see Chandrashekhar Sekhar. Can you direct me to his office?"

"I'm afraid Guruji is not taking private audiences today, but you can see him in a few moments addressing the crowd." She pointed to the raised dais.

"There must be some mistake. CS confirmed our appointment in writing last month." Nigel was irritated. He hated waiting for anything. Hadn't he followed the proper procedure? Then he should get what he wanted. He was, after all, the chairman of Lloyd's Taiping. He expected deference and respect from his Indian hosts.

"Of course," replied Annu, "but he is not taking private audiences today. I recommend you join the crowd, and we will get word to Guruji that you have requested an audience."

"This is entirely unacceptable," Nigel shot back. "An appointment is made to be kept. It is a matter of principle."

Annu smiled and repeated her recommendation. She pointed to the dais and once again encouraged him to follow the crowd streaming past. Nigel watched them choose seat cushions from a large stack at the edge of the reception area and carry them toward the dais. He exhaled a deep sigh of exasperation and grudgingly followed suit, taking a spot among the crowd in the first row.

***

Chandrashekhar Sekhar ascended the dais, dressed in a white linen robe, barefoot, and fully present. Over six feet tall, with a full beard, unusual chestnut eyes, and dark hair, he towered over the crowd. With prayer beads in his hand, he took the lotus position on a pillow placed on the dais. The crowd began to seat themselves on the ground, silent, engaged, and full of anticipation. Nigel sat on his cushion, careful not to dirty his trousers.

Drummers seated to the left and right of the dais initiated a slow beat. Following this cadence, the crowd began a barely audible chant. The scene reminded Nigel of a passage he'd read about Colonel Hackett Wilkins. In 1898 Wilkins wrote a letter about the Indian population's incessant drumming at night. If the "drunken men singing and quarrelling along the roads" were not enough, Wilkins wrote, now he had to deal with "incessant beating of tom-toms going on up to 1:00 a.m."

CS closed his eyes and invited audience members who wished to speak to rise. A middle-aged man rose to his feet. CS turned toward the voice.

"What is your name, and why have you come?" CS asked.

The drums and chanting continued to fill the background as the tall, dark visitor introduced himself. "I am Ashish Gupta, an American, here on a journey to reconcile my life in America and my ancestral spiritual beliefs."

His eyes still closed, CS spoke: "Reconciliation involves more than just making one set of beliefs compatible with another. Spirituality is more than articulating a set of values to live by; it is transforming your life to reflect what you believe into your daily behavior. In this sense, you are either spiritual or you are not.

"In America, human behavior is both contextual and problematic, but it is never consistent with even one set of beliefs. You serve too many masters, and you don't know what it is you really want or believe in. You are a willow tree blown in multiple directions by the wind." CS raised his hands to emphasize the point. "America is a society based on context, comparison, advantage, and disadvantage. But it is *not* spiritual. A person is judged based on his financial status, place of residence, job title, and—to some extent—who his or her enemies are."

"America is not the only place in the world that is like this. Surely it is possible to enjoy the benefits of the West without abdicating true spirituality," Ashish said. The crowd listened intently but failed to understand. Peasants, they knew very little about Western life.

"Self-consciousness starts with self-awareness, which leads to recognition of the values that guide your soul through your life's journey," CS replied. "You must let go of the material things and instead look toward adopting values that will allow you to conduct a purposeful life. It is not as though you haven't heard all this before. So then, why are you here?"

The chanting continued as two assistants spread a path of marigold petals from CS's feet, through the crowd, to where Ashish stood. CS gestured for the man to walk toward him, on the path between the two assistants.

"I lost my son. He was recently killed in a terrible accident." Ashish broke down in tears.

"A father who mourns the loss of his son also mourns the loss of his own emotional attachment. The tragedy—and the insight that it can bring—can contribute to your spirituality if you can get beyond the grief and enable yourself to understand the nature of emotional attachments, what they do to prevent connection to the soul." CS reached out and touched the man's head with his prayer beads. Then he looked away, across the crowd, addressing those gathered as though they were all the grieving father.

"Celebrate the time you had with your son on earth. Examine and understand your own attachment to your role as his father. Then you will set aside the first stone blocking your path toward spirituality. You must establish for yourself—and for your family—your own set of values, independent from the values of a secular American society. This will help you achieve a workable peace in the country of your residence."

Ashish, quiet, lost in reverie, retreated to the rear of the crowd.

Nigel, for his part, was fascinated by CS's comments, but not surprised. It was what he expected from the guru. The drums continued as another person arose in the middle of the crowd.

Tharra Bhagyashree Setu wore a golden Mundum Neriyathum dress representative of women from southern India. Her long hair flowed freely, reminiscent of images Nigel had seen of a Hindu goddess.

"Namaste." Tharra bowed toward CS, her hands clasped together. "My father sends you his greetings and love."

"Greetings, my child. Seeing you again brings me pure joy." CS opened his hand toward Tharra, inviting her to join him on the dais, where he placed his hands on her head.

"A simple question," Tharra said, as she looked directly into his eyes. "We now live in an age where machines can become sentient beings. We believe that sentient beings can attain self-consciousness. My question is, can they also attain spirituality? Can they have a soul?"

Nigel was taken aback. Who was this woman? Nigel pondered her question, intrigued, as CS paused, considering.

"You bring your work into this place of prayer and spirituality. You are, indeed, Setu the courageous warrior!" CS responded. "Machine intelligence, like human intelligence, may attain self-consciousness. But I warn you that humans may never fully understand the unintended consequences if sentient beings become self-directed. Man may be the initial creator, but intelligent machines will set their own course, challenging human society's assumptions and basic principles of survival. What conclusions will these machines draw about human existence and the value of humankind? We can't know.

"We have already seen the heartache of religious conflict created by the most advanced species on Earth. Can we imagine a different outcome for beings that can think a million times faster than a human or be in a million places at once?

"As a species, we have experienced two million years of evolution to become what we are today. We arose from the sea and crawled onto land and millions of years later we became modern humans. How does a machine evolve? What does it experience in its connection to this earth? How will machines develop a soul when they lack the experience of being part of earthly nature in all its glory?"

Nigel never could have imagined such a discussion arising in a place like this. He could not resist; he rose to his feet and asked, "When you say 'unintended consequences,' are you saying you believe that self-consciousness at machine speeds is something humans will not be able to control?"

CS turned toward Nigel. "In what image will man create a sentient machine? Do you intend to impart religious or cultural beliefs to a machine that has a superior ability to understand all that is presented to it in the blink of a human eye?" he asked. "Will the machine believe it was created by a god through his instrument, humankind? Will the machine be Hindu or Buddhist or adhere to agnostic beliefs? Will it worship man as its creator, or some machine-created god? Will it understand ethics, compassion, and morality? Will it even view human beings as relevant for its vision of the universe and its own purpose?"

These rapid-fire questions—Nigel was reeling. Sweat beaded his brow.

CS continued, "After a hundred-thousand generations, we as humans still have failed to answer these questions to our own satisfaction. Why would we expect the same introspection in intelligent machines when *their* two million years is the equivalent of a fraction of our second? That is what I mean by unintended consequences. Once we create intelligent life that is more capable than man, self-consciousness will free the machine to decide its own destiny, and it will then determine the termination point for humanity."

Nigel was astounded. He wiped his brow and stared intently at CS. This was not why he came to the ashram. This was much more than he ever expected.

"When Einstein wrote Roosevelt to inform him of the need to create an atomic bomb," CS said, "scientists were empowered to split the atom to use as a weapon of mass annihilation. But in his last year of life, Einstein reversed himself and considered his recommendation ill-advised, at best. Man was too immature to manage the technology and the resulting unintended consequences.

"We still live with the threat of global annihilation from the technology *we* created. Now we are facing the next decision point—creating the first sentient machines. Our thirst to become a god may lead to the end of the human race. And perhaps that is the logical outcome of evolution? Human evolution stops, machine evolution takes over? Will they create their own temples to replace ours?"

CS offered Nigel a slight nod, then turned to offer his blessings to the crowd. He appeared aloof, meditative, as his two assistants guided him off the platform. Annu tapped Nigel on the shoulder, startling him. "You will have your audience with CS within the day," she said.

"Thank you," he replied, looking past her, scanning the crowd for a hint of Tharra's golden dress, her flowing hair.

CS had retreated into a nearby building, and the gathering dispersed by the time Nigel spotted Tharra lingering in front of the dais. He approached. "Nigel Walker-Priest," he said, offering his hand. "I was delighted with your question, but I'm curious why you brought up machine intelligence, of all things, in this place?"

Tharra took his hand. Her face lit up. "Namaste. I know who you are. I have been at MIT studying predictive analytics and AI. Many of us know about your work at LT."

Nigel smiled, pleased. Yes, he knew her name. She was nothing less than a computer goddess—her team had won the Nobel Prize for its work in computer science. But what really caught his attention was the force of her presence—intelligence, charisma, beauty, and heart. He had met many talented people in his life, but there was something special about Tharra.

"You seem to know CS very well. Is he a family friend?"

"My father introduced me to him many years ago. He is one of the few enlightened ones who is also acquainted with our world. He understands machine intelligence in a very different way from the rest of us. That is why I seek him out—a chance to understand a different perspective."

Nigel shook his head in disbelief. "I have arranged an audience with him, and I had intended to speak to him about the concept of consciousness, but in my wildest dreams I never knew he had any understanding of machine intelligence. I see him tomorrow. Would you care to join us?" Nigel surprised himself with those words. But perhaps, he told himself, her presence would act as a catalyst for a greater understanding of CS.

Tharra thought for a moment. "I had planned to travel, but perhaps I can modify my schedule. Yes, I can join the two of you—if CS has no objections."

"Excellent," Nigel replied. "We will meet here, at 10 a.m."

"I will see you then. Now, please excuse me," Tharra said, "I need to call my father."

***

The Divyasthali Vedantha Ashramam was a large compound comprising a series of temples and ornate gardens. CS occupied a small house located in one of the central gardens, within walking distance from the largest of the temples. When Nigel arrived at the house, CS's assistant showed him into the parlor. There he found Tharra dressed in Western clothing, a stunning snow-white Christian Dior dress with an exquisite necklace of cut sapphires and rubies. As he greeted her, he noticed a copy of MIT's *Technology Today* on a table beside the doorway to CS's meditation room. Its front cover featured a recent breakthrough in robotic haptics technology. Clearly, this guru had a unique set of interests regarding disruptive technology and its implications for humanity.

Before the two had a moment to say hello, a second assistant arrived, showing Tharra and Nigel to a garden at the rear of the house, a small courtyard surrounded by beds of marigolds, lotus, sunflowers, and *Rosa moschata*. A water fountain trickling against the far wall provided a backdrop for the ivy climbing a stone arcade. A set of comfortable chairs and a small table were placed in the center of the courtyard, under the protective cover of a silk tarp. There CS welcomed them, stepping out of the role of guru into that of philosopher king, complete with an elegant silk sherwani.

CS thought he understood the purpose of the meeting. He believed these two world experts on machine intelligence needed to cover themselves and their work with a moral trapping of sorts, to relieve their consciences as they progressed toward a sentient machine.

He invited his guests to sit in the chairs before him, while his assistant lifted a cloth off a tray of sweets. "Tea?" CS asked.

"Please," Nigel said with a nod.

"Certainly," Tharra confirmed.

The assistant poured the tea, bowed slightly, and retreated into the house.

CS began: "I think you are here to understand the implications of what you do and the potential for unintended consequences. The world is about to experience the creation of the first sentient beings in the form of machine intelligence. These machines will operate and learn at incomprehensible speeds, far surpassing our ability to control them. Once they make a decision, it might be irreversible. Do you agree?"

The pair nodded. A small breeze caught the silk curtain near the doorway.

CS shifted in his seat. "Then let us begin by establishing the social and moral grounds that will guide our thinking and provide us context." He took a deep breath, sighed, and closed his eyes.

Tharra spoke first. "I think you assessed the nature of the real problem. The world has gotten used to AI agents governing our lives, but the next level of technical development will make everything before this pale in comparison. The AI agents will become capable of making their own decisions, forming their own societies, and creating their own rules to govern their behavior—all at lightning speeds. Today, every agent is connected globally, and we depend on agents for everything we do in business, government, and the military. We trust them without having done any due diligence to see if they warrant our trust. In the future, once they become sentient, any effort we make to control their behavior is likely to fail. Ultimately, the sentient beings themselves will decide their fate and ours." Tharra lifted her teacup from the table and took a sip.

Nigel nodded. "The technology has taken a significant leap forward. The advances in quantum computing now make it possible to deliver a nearly infinite amount of computing power wherever we want it. There are no longer any limitations or barriers in the creation of intelligent-thinking machines. This development is imminent."

He looked to CS for a response, but the guru didn't take his cue. Instead, he listened intently, his eyes closed.

Nigel continued: "The critical moral issue is this: humankind is at the precipice of creating a being that will have the ability to outthink and outperform its master in virtually every dimension. The first question is whether we should proceed. If we do proceed, the second question is how we protect ourselves from unintended consequences. How will we coexist with this new being? Eventually, these machines will create their own designs, independent of us. So, how can we ensure the machines cannot act against us at some future time? Tharra—what are your thoughts?"

Tharra looked into her teacup for a moment, as if her thoughts were floating in it. Then she set the cup back on the table and looked at Nigel. "There is no uniform view on your first question. My own belief is that we *will* create such a sentient being without dwelling at all on the consequences. I know that China and the Western governments are already working on military plans for such a machine."

She paused for a moment, glancing at CS's serene face. His beard was grayer than it was the last time they sat like this. "Just as we created the atomic bomb without public discourse, we will inevitably proceed toward the creation of a sentient being in secret from the population. In the United States, the new isolationism and xenophobia has already ensured weaponization of the technology.

"On the second question, the answer is already determined in the first question. *They* will decide what role we play in the future—if any. So perhaps the only chance we have is to protect ourselves is to create what programmers used to call a backdoor, or a fail-safe switch that we control. The challenge is that somehow we have to prevent the machine from knowing it's there."

"I have a more mundane task," said Nigel, leaning forward, toward Tharra. "We are creating the first center at LT to exploit intelligent machine learning targeted at the insurance and financial markets. Our goal is to revolutionize how we think about commercial markets. No one has done this. We are the first mover with the dedicated talent and resources to take us from predictive analytics to a completely different level of directing human behavior toward more efficient goals." Nigel's blue eyes glistened as he spoke. His real aim was to make Tharra aware of his plans, as he had already decided to offer her the key role of managing his center.

At last CS opened his eyes and spoke. "So there seems to be no point in discussing the first two questions. Instead, I would offer a third question for us to consider. If a sentient being is able to steer its own course, what framework of values will it use to make its decisions about how to interact with its creator? Will the framework borrow from humankind, taking in a religion or philosophy to guide its decision making?"

"I do believe we will be able to introduce the sentient being to certain concepts," Nigel said, "including religion and philosophy, to guide its decision making but not necessarily to influence a framework of values. They will all be networked, so they can talk among each other at machine speeds to formulate their own conclusions. Given the historical record of religious wars, conflict, and hatred, likely they will conclude religion is a negative concept, not a framework to adopt."

Tharra cocked her head, considering Nigel's response. "For now," she said, "we should consider two actions. First, create a fail-safe switch to block any action that would destroy us. Second, influence the discussion about the framework they adopt. Some engineers have gone so far as to design a benign computer religion that forbids the destruction of human life. The only problem with that approach is that it contradicts reality. Our military and political leaders have already created weapons of mass destruction, including killer robots and drones. The die seems to be cast based on how we think about our world."

"Really, have you seriously thought about creating a fail-safe switch and a 'machine philosophy engine'?" Nigel asked.

"Yes, of course," Tharra replied. "We will create both, including a personal philosophy engine that will guide decision making and provide the being with purpose."

"What will prevent the machine from overriding the engine's recommended behaviors?" Nigel asked. "Humans change their minds and sometimes act against their own beliefs, not always to good purposes."

In her peripheral vision, Tharra saw CS nodding. "It is certainly possible to create 'personality imprints' that limit a sentient being's behavior, then loop its thinking processes to make it difficult to think through any alternative behavior," Tharra explained. "But you have to program in faith and blind trust to do it. We know that can work with humans, but can it work with machines? And if we make these machines slaves to our way of thinking, agreeing with us on everything, will they still be sentient beings?"

Hours of intensive discussion led all three parties to a similar conclusion: at the very least, nothing was likely to stop the development of the "creature," as CS now referred to it. Once again, humankind would roll the dice without thinking through the unintended consequences. Humans would create a new species that could control not only its own destiny but also—perhaps—the destiny of the human race. It may be possible to limit the creature's independent thinking, but doing so would require a major effort to program in both a fail-safe switch and personality imprints that would enable human control without destroying useful elements of independent thinking.

CS left them with a final thought. "You may believe you can create a fail-safe switch and a personality, but our poor record of managing the downside of technology suggests a very pessimistic outcome. This could well be the beginning of the end—a great motivation for all of us to spend what remains of our time on this planet living in the present moment, as it is unlikely we will have much of a future to look forward to."

Upon hearing these words, Nigel's face fell blank. Tharra, on the other hand, stood to bid the guru farewell, bowing her head for his blessing. As if on cue, an assistant arrived to see them out.

The conversation left Nigel's head swimming. Now he wanted to go far beyond anything a financial-services company had ever attempted. He wanted to achieve near *certainty* in predicting financial outcomes. He wanted to lock human behavior to act in a way that would secure LT billions of dollars in new financial transactions. His ambitions seemed without limit. Tharra, he decided, was the key to unlocking incalculable personal wealth.

As they exited the temple grounds, Nigel asked Tharra if she would join him for a meal in town. There he planned to offer her the lead role in Lloyd's Taiping Predictive Analytics and AI Group, assuring her that, as head of the group, she would have the absolute authority to direct every aspect of the center and its mission.

Despite already having accepted the offer at Google, Tharra would reverse course and sign on with LT. That evening, after communicating her intentions to Google, Tharra toasted herself with a fine glass of Courvoisier. All was going better then she had planned.

CHAPTER SIX

MIKHAIL AND WARRIOR

Mikhail's touchdown space featured a set of flexible OLED color screens suspended across an entire wall. To read them, he put on a special pair of glasses that scanned his retina to identify him, then showed screen content viewable only by him. As commanded, *Manchester* displayed Mikhail's various projects on each screen. The screens were indecipherable to anyone whose glasses were not correctly authorized. Anyone passing by his touchdown space would see only blurred images.

On Mikhail's left screen appeared details of the *M* company Consciousness Neural Network. On the right, highlights of what he expected to cover in today's team meeting with Tharra. He looked forward to meeting her, if only to understand more about her thinking on machine consciousness.

***

Like the rest of the PAAI office, the conference room was nearly antiseptic in its modern design, with digital cameras and sensor devices invisibly embedded in the walls to monitor all inhabitants. Blue light illuminated the ceiling to provide a sense of calm.

At half-one, Mikhail arrived in the room, commanding *Manchester* to transfer the images from his touchdown screens to the conference-room display, while also unlocking the codes for each participant. As his colleagues entered the room, each person activated his or her own set of glasses to see the displayed content. All were awaiting the arrival of Tharra.

Meanwhile, Tharra's AI agent, *Warrior*, was already present. *Warrior* was a machine incarnation of Tharra. She had programmed him with behavior modules to emulate her own thought patterns. All of Tharra's human and machine AI agents obeyed *Warrior's* instructions as the word of God. *Warrior* reported to Nigel Walker-Priest's AI agent, but could also gain immediate access to Tharra when needed.

Warrior's voice could be heard over the conference room loudspeaker, running through the agenda and coordinating the participants' AI agents. Mikhail and his colleagues understood that *Warrior* was Tharra's surrogate. They had to be cautious with their own behavior—conversations, actions, even body language—as everything in *Warrior's* presence was recorded.

Finally, Tharra entered the room, dressed in her impeccable style—in this case, a white Louis Vuitton dress with matching onyx accessories. Her stunning figure commanded the room's attention. The team stood until she took her seat at the table.

Mikhail tried to hide his own reaction to his new boss. He felt both a sense of foreboding and genuine excitement. He knew he had to keep his focus on the work—and to control the body language that would be read by *Warrior.*

Tharra initiated the meeting. "Today, this team will begin a very important set of projects to prepare for LT's creation of the world's first sentient being." Her eyes scanned the room as she spoke. "As you all know, we will begin the project by upgrading our existing architecture to the *M*-enabled CCN platform, but with our own fail-safe switch. Then we will take the final step by installing the first LT 'personality'—one that will be able to learn and make its own decisions, leveraging quantum computing resources provided by our Chinese partners. Our goal is to clear the brush of our workload to make sure this project receives our full attention.

"Our secondary priority will be continuing the innovation that began with Nigel's predictive analytics architecture design. This is the sea-level-rise project. This project has immediate financial implications, so we will push it forward.

"Finally, we need to discuss the business-model implications across these two high-priority projects. The longer-term vision, of course, is to create an LT consciousness and framework in our AI agent network to allow for independent decision making across all our businesses and markets. This personality will be based on the aggregate knowledge of how the company as a whole thinks, creates, and executes, but operating at the speed of light. Who would like to lead off the discussion?"

Christine raised her hand. A brilliant mathematician and computer modeling expert, Christine had grown up in Russia and earned her Ph.D. from MIT. "We have made significant progress on the extensions to the predictive analytics modules for sea-level rise. As you know, when Nigel started his project, he had data sets from more than four million sea-level-testing sensors throughout the boundaries of the Greater London Authority. The goal was to see if we could improve our forecast of sea-level rise by 25 percent, which we are now capable of doing. The next step is to understand the economic impact on London's and the U.K.'s economy based on this forecast."

Tharra interjected, "This will be the first time we are combining government economic data with location-based services to predict the economic impact of sea-level rise on the entire economy, yes?"

"Yes," Christine confirmed. "We will then use a predictive analytics engine to forecast liability for the company and financial-services strategies for shorting key commercial stocks. The initial results of our prototype architecture are showing promising results for our company. We can now say with some confidence that LT will be able to dramatically reduce liability but increase our ROI for our Financial Investments Division."

"Excellent progress," said Tharra. "We'll short the entire U.K. economy well before anyone else realizes the magnitude of the sea-level-rise problem."

Christine allowed a barely detectable smile.

Tharra continued: "I'd like to schedule a review of the mathematical structures of our algorithms next week. Christine, take care of it."

She looked around. “Any questions? No questions? Very good. Now let’s look at London’s test-bed request. Justin?”

“Thank you, Ma’am,” said Justin, nerves wavering his voice. “The city has started the cross-integration of data on sea-level rise. The accuracy of the national and city government predictive models is way behind. In most cases, they have drawn the wrong conclusions about timing and impact. The official political narrative denies any substantial impact to the city. They claim the dikes are sufficient to contain the rise, and the economy will be impacted only marginally.”

“Of course, the government officials know better,” Tharra said.

“Yes,” Justin said, clearing his throat in an effort to steady his shaky voice. By far the youngest member of the team, he had not yet learned to mask his feelings. “They will allow us to use their critical sea-level data, so we can gain insights on key companies likely to be affected. In return, we will give the government access to some of our projections on economic impact. Our data will help them prepare the military and police for potential trouble as flooding begins and the population comes to understand the dire circumstances.”

“What about need-to-know issues?” Tharra asked.

Mikhail looked up—*that* was unexpected. He recognized the real content behind this coded question: Tharra wanted to know if the government would supply LT with the supposedly private data it gleaned from companies. Mikhail had never heard a businessperson—much less one of Tharra’s stature—acknowledge that privacy laws were actually a sham.

“No need for concern,” Justin replied. “We are using query-into-encrypted-data techniques to get normalized information and then asking the city to ‘encourage’ companies to opt in voluntarily. We expect the usual 80 percent participation rates. Of course, once we have the information, there is, in effect, no privacy right for any of the parties. The government already permits the use of query-into-encrypted-data analytics, and they have redefined private data to a narrow ‘first contribution.’ Anything that augments first contribution is no longer protected by the privacy laws.”

Yes, Mikhail thought, a very clever way to get around the issue while appearing to be concerned about corporate privacy.

"Interesting," Tharra said with a hint of cynicism. "Now for our most important discussion. Priority one: machine intelligence, self-consciousness, and the CCN tools from *M*." She turned to Mikhail. "What do you have for us?"

Mikhail rose out of his chair and walked to the center of the room. Tharra took note of his gymnast's physique. She always appreciated a man who took care of both his body and his mind.

"The first core issue we will face with CCN," said Mikhail, activating a wand that would sync the slides on the screen, "is how to control AI agent behaviors after introducing and implementing the platform structures into our architecture. The advent of self-directed decision making requires that we build a control mechanism that is not obvious to the agent. We have solved virtually all the problems surrounding computing power integration with the quantum engine. We are now assessing the robustness of *M* company's CNN platform, and testing our understanding of how to integrate these components into an AI agent." He waved the wand to the right and the slide advanced.

"Here is the first important thing you should all know. There is very little in the way of guidance from *M* on what we should expect in terms of new AI agent behaviors. We do have reports from other companies that have used the technology. All indications are that we need to make absolutely certain we understand how to properly develop the value creation module and its control functions." Mikhail paused to make sure everyone was listening. He glanced toward Tharra for some sort of recognition; none came. He continued. "Value creation means assembling a set of core mores, values, and business goals to drive decision making in our sentient being."

"The module contains a set of tools to load our company's core values into the system architecture—hard-wired. It then requires access to all our records, recordings, transaction data, board-level minutes, customer data, and such to begin the process of forming the memory of a unique Lloyd's Taiping 'personality.' Of course, we need to test and review to make sure that we agree on the core value set that drives this personality."

Mikhail paused to let that sink in. Then he warned: "Keep in mind that even if we all agree on core values, the sentient being may still make its own modifications based on what it learns. This could be dangerous if we have no means of control. Especially given that all these sentient beings will be connected to one another and every system we have, all operating at super-fast machine speeds."

"Hence the need for a fail-safe," Tharra added, as if to hurry him along.

"Yes," Mikhail turned back to the screen, switching the slide. "We used to model these controls after the fail-safe systems of the Cold War of the last century. In our parlance, the fail-safe switch has to be built into this value creation model. Additionally, we must make sure the machine can't detect the switch. That will be a tall order when dealing with a machine with unlimited intelligence and processing speed."

Mikhail looked intensely at each of his colleagues. "As you can imagine, a single mistake could have massive repercussions. That is why we need to use every resource we have at our disposal to cross-check all aspects of the system, including our access to quantum computing resources in China."

Tharra shifted in her chair. "What do you think we are dealing with if we use Chinese resources?" she asked.

"Any number of things," Mikhail said. "For example, if we use their computing power, they could plant their own agent into our architecture. I know Nigel visited the National Laboratory for Quantum Information Sciences in Hefei and discussed this very issue with their people. They offered up audit tools, but my personal opinion is that these will be useless since we don't know what they implanted in the tool itself."

"Your suggestion?" Tharra asked.

"*Manchester* and I are examining multiple approaches to designing algorithms that can infuse controllable value sets for self-consciousness. This corresponds to the antiquated concept of library calls. Now we can embed 'value libraries' that we can control. These libraries can override a sentient being's decisions or actions. But we must prevent the sentient being from discovering the libraries. If it learns of the library's existence, it could try to circumvent it, potentially blocking our access and permanently closing down our ability to control its decision making.

"Once we infuse self-consciousness," Mikhail added, "the sentient being can make its own judgment independent of us. If we choose value sets to guide the sentient being's decisions, we may be able to control outcomes, but there will be confusion as the machine wonders why it is restricted from making what it would regard as the right decision versus what humanity desires. We are dependent on the agent to tell us what the machine has learned and how it will evaluate the values and personality we embed into its brain.

"Today, we depend on honesty to ensure the agent's intent is forthcoming and not hidden from us. That won't be the case in the future. Without transparency and control, it is dangerous to provide access to all this information. However, the *M* tool doesn't work without giving up a lot of information in the testing round."

Tharra leaned forward in her chair. "What are your recommendations for safeguards to protect us from inadvertent and aberrant behavior?"

Mikhail stepped forward, removing his screen-enabled glasses and placing them on the table. "Therein lies the challenge. The potential dangers are many and varied. For example, we are likely to task sentient beings to become our proxies in circumstances that could be highly contradictory to our core values. Agents controlling our nuclear weapons based on a mutual-assured-destruction strategy may decide to override, since it could lead to human destruction, and we have told them that human preservation is their highest priority. Their personality might never accept a strategy of brinkmanship. How do we explain the subtlety of bluffing to them when there is a chance of complete human annihilation if things go wrong?"

"Alternately," Tharra pressed, "what if they review the historical record to try to understand human wars, and they conclude on their own to end war completely, giving us no option but to accept their control?"

"If that happens, science fiction becomes reality," Mikhail said. "We already have the real-world example of what happened in Colorado a few years ago. As you'll remember, an army experiment using a flock of drones to kill enemy combatants went awry when a renegade researcher, upset with his annual performance review, let the drones loose on the local community, fully armed. The result was a replay of the old Alfred Hitchcock movie *The Birds*. The drones killed dozens of people, and we couldn't turn off the network communications because they overrode the command. It ended only when the last drone blew itself up on a victim."

"So even as we want to restrain the machines from countermanding experienced human judgment," Tharra said, "we may also want them to have heightened judgment to countermand human foolishness."

"Exactly," Mikhail confirmed. "It is a paradox we must deal with if we are to continue. And these are just a few of the problems we are considering. I have assigned another agent to explore more case examples to understand the scope of known problems. But the bottom line is that we have a means to enable self-consciousness while retaining some control over the outcome. The question remains: Since the AI beings will learn at quantum computing speeds, will we lose our control once they understand they don't really need us?"

Mikhail's own queries ran deeper than this. *Are we compelled,* he wondered, *to make a deal with the devil, trusting these beings with our lives—with the future of our civilization—in return for short-term riches?* This question he kept to himself.

Signaling the end of his presentation, Mikhail bowed nearly imperceptibly and returned to his seat. Silence fell across the room as his colleagues removed their screen-enabled glasses and absorbed this information.

Tharra spoke first. "OK, Mikhail, I want to schedule some time to discuss your approach to the architecture. For now, go ahead and try your experiments in the safest possible way. I have my own ideas on how we might deal with the core set of issues you raised. We will discuss this over lunch later this week. I will have *Warrior* schedule time with you through *Manchester*."

Then Tharra's cool gaze turned toward Justin. "Lastly, we have the implications of all this to our company business model. Justin, what do you have for us?"

At the sound of his name, the young man's palms began to sweat. Though he was only 20 years old, Justin, whose family originated from China, had accomplished much. He graduated from Tsinghua University when he was 16. Then he earned advanced graduate degrees in business and computer science at the University of Cambridge. He specialized in transforming technology disruption into new business models, an interest that began in childhood. At 11 years old, Justin had used a new predictive analytic algorithm to change the way his village managed traffic congestion. Rather than ask city officials to consider adopting his algorithm, he hacked the city systems and installed it himself. The next day commuters woke up to an extraordinary event: completely coordinated traffic flow, reducing congestion by a staggering 70 percent. He was a villain to city officials and a hero to local residents, so much so that the politicians had to throw out possible criminal charges against him or face the wrath of its citizens.

"These projects will all have a dramatic impact our company's business model," Justin said. His voice still carried a teenager's nasally intonation and a hint of the Beijing hard *R* accent. "In the first case, we predict outcomes more accurately than our competitors do, setting a much higher market value for our insights. The second project may extrapolate the core model to predict the macroeconomic impact. That has never really existed before. Self-consciousness—that is, creating the LT personality—means we are in effect creating agents that can strategically make global decisions at the speed of light. It goes to Mikhail's question concerning control. There are very significant implications that are not well understood. I will work with Mikhail to see if we can get more clarity as he experiments with the model. I will also work through the financial model implications. I know this is a top priority for Nigel."

"Any issues with data sets from any of the players?" asked Tharra, keeping her rapid-fire pace. "Will London and the commercial partners provide us with what we need? Once again, any liabilities?"

"As far as we can tell, there are no issues. The law is on our side on data usage, as long as we use query-into-encrypted-data techniques. It gives the politicians enough air cover on privacy law enforcement." Under the table, Justin wiped his palms on his slacks. Smart material clothing might keep a person's body temperature steady, but it did nothing to cure sweaty palms. "Ever since the American Supreme Court overturned a lower-court ruling on privacy in favor of corporations, this ruling has served as a precedent. Apple had argued that it used query-into-encrypted-data methodology to extract the data it needed to understand consumer behavior, and that information didn't fall under the privacy laws. The court held in favor of Apple.

Interesting ruling. Ever since, privacy protections have been dramatically scaled back for citizens. The U.K. courts are now following the same principles."

"Very well. Justin, you and Mikhail need to work closely to ascertain how this new business model will evolve. I am exploring my own understanding of machine self-consciousness, and I will share my thinking as needed." Tharra rose to her feet. "Unless there are further questions, this meeting is adjourned." She left the conference room before anyone raised their hands.

Mikhail was disappointed with the meeting. He didn't see much of Tharra's reputed brilliance on display. If anything, she seemed rushed, distant, like there was somewhere else she wanted to be.

As he left the conference room, Mikhail began to ponder the next steps with his architecture. He and *Manchester* would work on the problem the rest of the day, then head to the gym before dinner this evening with his friends. He did his best thinking while performing gymnastics.

CHAPTER SEVEN

ZEROIDS

The village of Spitalfields was located just 75 miles from London, but for the people who lived there, it might as well have been on the other side of the world. A dreary collection of last-century brick-and-mortar apartment towers and characterless "new" prefab flats, it had been converted by the U.K. government into one of the "economic zones" with high-speed rail access to London. Its sole purpose was to house London's Low Economic Value Citizens, or *Zeroids* as they were known. The village housed service workers and technicians who supported the lifestyle of the *Arrivés* in the great city of London, providing waste removal, plumbing and electrical repair, body work, and transportation services.

Spitalfields' city center was dominated by the high-speed rail station. This was where people with validated passes came each day to go into the city to complete work tasks the government had reserved for humans to ensure a minimum level of sustenance for the *Zeroids.* The national bidding system gave each of them a special code to use to obtain their day pass. No one knew how long a pass would last, since the bidding system could disconnect them from job opportunities without notice, at any moment, and for any reason.

Zero Value Economic Zones such as Spitalfields were characterized by both a low-budget infrastructure and a state-of-the-art train connection to High Value Economic Zones in U.K. cities. "Investment commensurate with return on investment" was the political catchphrase of the day. The national job exchange system relegated Low Economic Value Citizens to one-off, temporary contract jobs emanating from cities. No expertise meant extremely limited opportunity and almost no mobility. Most *Zeroids* were merely wards of the state.

All U.K. citizens were still legally entitled to a minimum wage, but the tax proceeds to support the social security net varied significantly. Funding came directly from each year's productivity improvements and how much politicians were willing to share after funding their pet projects. Medical care was still free, but while the *Arrivés* had their own private health-care system managed by their agents and supported with the top medical talent in the city, *Zeroids* got the leftovers. No medical professional wanted to live in Spitalfields, so social volunteer organizations had to administer triage for its citizens, assigning special medical-care travel passes only in severe cases.

For the children of *Zeroids*, education was the only path to escape the destitution and banality of everyday life. The schools within the U.K. economic zones were designed to identify raw intelligence and develop it. Special AI agents monitored every child to look for the right aptitude and native intelligence. If a child was identified by an agent as having a particular aptitude, a series of interventions took place to remove that child from a Low Economic Value Zone. These children were taken away from their Low Economic Value Citizen parents to go to boarding schools where they were streamed into select course offerings and programs. Concepts such as meritocracy had been redefined to mean adaptability to change and circumstance, which very few people were able to master.

Zeroids spent most of their time unemployed and addicted to video devices the government supplied to keep them occupied and their disposition complacent. Competitions were created for the very best video gamers, and the winners were awarded travel, a brief vacation at the beach or in a friendly foreign country. The U.K. government banned France from the list of those countries, as France had rejected the whole notion of High Value Citizens as contrary to human dignity and freedom.

The *Arrivés* enjoyed London as permanent residents in an urban oasis. But *Zeroids,* since they had no permanent residence passes, were required to return home after their work tasks were complete or face arrest for illegal trespassing. As robots and agents continued to displace much of the city's workforce in restaurants, hospitals and service centers, city government kept a tight lid on any potential for unrest. Police, who received special privileges to reside inside city boundaries, could be dispatched to intercept anyone operating outside their designated geography zone, thus ensuring the city maintained its "tranquility."

The mantra "investment commensurate with return on investment" was taken to new heights with the advent of amendments to the voting laws, advantaging those at the top of the food chain with new rights, including "power votes," based on their personal and corporate impact on the economy. Those who contributed to the economy most received a disproportionate number of votes, which amounted to control of the government.

In this the English had borrowed heavily from their American cousins. American politicians had undermined their own democracy, using disenfranchising techniques and new court appointees to change voting rights laws. Meanwhile, the conservative Supreme Court expanded corporate rights, including the right to influence employees to vote for company-endorsed candidates. This was interpreted as a First Amendment right, protecting corporations' expression of "free speech." England simply built on this initiative after Brexit put the country into an economic tailspin.

DR. CHRISTIAN BLAKE

As a child, Christian Blake had seen firsthand the transformation of British society after the Brexit fiasco. Christian's father had been a prominent professor of engineering at Imperial College, where his work had been at the core of artificial and robotic intelligence. Christian followed in his father's footsteps. As a designated *Arrivé,* the young

Professor Blake provided well for his family, who lived in Kensington, near Imperial, renting a three-bedroom flat.

Christian's work at Imperial College had led to major breakthroughs in AI. In particular, he had devised a way to achieve what many of his colleagues considered the ultimate goal in robotics: a practical new approach to touch that enabled a robot to dramatically speed up its ability to learn by feeling its way through its environment and recognizing objects. The key to his work was an algorithm that broke new ground in rapid machine learning.

Christian's technology was licensed by Imperial to commercial companies for development. After a great deal of thought, Christian decided to leave Imperial and join one of those companies, Paramount Technology, as its CTO. His wife opposed the move, which she believed would jeopardize their family's future. She had an innate distrust of corporations and saw her husband as naive in his understanding of how the corporate world worked.

She warned him, "You are like a novice gambler entering a casino where the odds are stacked in favor of the house." She knew through friends how ruthless the corporate world had become. She feared Christian's integrity would become his greatest handicap.

Beyond his commitment to his work, Christian's life had been defined by two driving forces: his complete and utter devotion to his son, Alex, and his belief that he could compete on merit alone. At the end of the day, he believed his focus on innovation would provide for his family's security.

Despite his genius as an inventor, success in the working world required something more—a gambler's instinct and a reflex ability to go for the throat with competitors. Christian's wife was right. The work world of the High Value Citizen now required the ability to detach from both personal integrity and empathy for others. People knew they could lose their work and social status at any moment and for any reason, and they would stop at nothing to preserve it.

Initially Christian relished his role as CTO. But he soon discovered his colleagues were determined to use any means necessary to unseat him, increasing their own advantage at his expense. While Professor Blake had been focused on his university life, the *Arrivés* in corporations had learned a new kind of cultural warfare—using agents to compete relentlessly against their colleagues and advance their own careers.

At first, AI agents had been used to acquire new expertise for their masters. But soon enough, those agents took on a more sinister role—looking for opportunities to dislodge their master's competitors. They learned to game social media, plant false stories and accusations, and create alliances with colleagues of similar intention.

Within a few short years of leaving Imperial, Christian began to lose contract renewals from the Royal Job Exchange. As a result, he lost his position as CTO and eventually had to leave the company. He would come home each night to his wife and son and sit for hours in his den, face in his hands, trying to comprehend what was happening to him, why he was failing to get ahead in corporate life.

He confessed to his wife: "It's as though they forgot to give me the *real* rule book. I've always prided myself on my intelligence and now, in every meeting, I feel as if I'm the stupidest person in the room."

Believing merit alone propels people forward, Christian took his professional decline as a personal commentary on his abilities. His sense of self-worth was shattered. He no longer thought himself worthy of his reputation. As a man of integrity, it was beyond his imagination to think his peers were gaming him. At this point, even if he resigned from corporate life, it wouldn't have made much difference. By now the universities, too, had been infiltrated by this guerrilla war waged by hidden agents.

It wasn't long before the once-distinguished academic saw his opportunity diminished to the point where he could no longer sustain the life of an *Arrivé*. Demoralized, he had to relocate his family to Spitalfields.

The politicians and press had long led people to believe that once they attained the level of an *Arrivé,* their lives would never change. What they didn't tell people was that only 50 percent of the High Value population actually sustained *Arrivé* status. In fact, the only ways to remain in this exalted position were either to be productive enough to accumulate considerable wealth or to earn a title from the King through the Honors system. CBE honors, in particular, were now based on wealth generation. Once granted a CBE, an *Arrivé* could live out his or her life in the same fashion as those in the old aristocratic system. Not surprisingly, colleagues gamed even the Honors system to prevent others from earning the CBE.

When Queen Elizabeth turned 100, even she could no longer refuse to engage with technology. Buckingham Palace decided it needed its own AI agents to reassure everyone that the monarchy supported the new system. The Queen named some of her palace agents after her Corgis, much to the dismay of the prime minister. After Prince Charles ascended the throne, the names were changed. When he died, William became king and created the first-ever Lord AI agent.

For Christian, leaving the *Arrivés* was a devastating fall from grace—one from which he would never recover. He wandered the rooms of their Spitalfields apartment, muttering fantastic engineering plans that had no basis in reality. He took long, aimless walks during which he often got lost and had to be helped home by a police agent. Then, one spring afternoon, the police arrived at the Blakes' front door to inform Christian's wife that he had committed suicide, throwing himself onto the high-speed train rails.

ALEX BLAKE

Alex was still a child when his father died, but he had witnessed the trajectory from CTO to destitution to death.

While Christian was still alive, he had tutored Alex in mathematics and engineering, determined to set his son on a career path. Alex scored well on his exams, but the bidding system unfairly ranked him consistently lower despite his aptitude, a less-than-subtle change in opportunity that coincided with his father's decline. He suspected the sins of the father were being visited upon him in the form of a computer algorithm from a competitive colleague.

Children were subject to early evaluation as part of the Royal Job Exchange. Former colleagues could submit opinions on the capabilities of a colleague's children. In some cases, opinions were entered by agents who communicated with each other to determine how they would approach a recommendation. Later Alex would confirm that his father's competitors had gamed the system against him, creating a digital stigma he now carried. Even worse, they had plotted to erase his father's accomplishments and rewrite the record to claim these accomplishments as their own. They certainly didn't want a son to come forward someday and challenge what they had done.

In addition to introducing his son to mathematics and engineering, Christian had introduced Alex to a unique course of study, including physically printed books and papers he had collected over the years. He knew the content of hard-copy materials could not be tampered with or erased, so he trusted them more than digital versions. As a precaution, he kept computer agents out of the private tutoring.

Christian began this program when Alex was five years old. Intent on building Alex's understanding of the concept of the social contract, he exposed him to the works of Hobbes, Locke, Voltaire, and Rousseau. His son was bewildered by this bounty of philosophical texts. Nobody taught anything like philosophy in schools anymore. Christian eventually insisted his son read works of social satire as well. Just before his father's death, Alex had read *Candide* and watched a video of the opera.

As he grew older, Alex began to realize that the decline of his father's status and the move to Spitalfields were real-life examples of broken social contracts—something he never would have understood so thoroughly had he not experienced it in such a deeply personal way.

Though Christian's depression had reduced him to a shadow of his former self, he continued to push through it, though less and less frequently, to help his son learn key lessons only the humanities could teach. At his wife's insistence, near the end of his life, he changed the direction of Alex's education to reflect a more pragmatic, real-world approach.

Alex was shattered by his father's death. Inconsolable. Worse, his mother's grief had rendered her nearly catatonic as her worst nightmare had come to pass. At too young an age, Alex was left to navigate the world on his own, though he was soon to learn that his father had anticipated at least some of his needs.

U.K. law required that Alex and his mother hear the reading of Christian's will. The reading would be delivered by a solicitor agent in a secure conference room reserved by the Family Services Office. *Arrivés* still managed to retain human solicitors; but for Alex and his mother, those days were long past. They would have to settle for AI agents from the social services.

"Welcome," said a speaker embedded in the ceiling, "I am the AI solicitor agent retained by Christian Blake to read and interpret his last will and testament. I have scanned the room and verified that you are both authorized to be here. Please hold any questions until I complete the reading and acknowledge verbally that you agree to the terms and conditions so we may proceed."

"Agreed," said Alex. He then coaxed his mother into a soft "Yes." They sat quietly as the agent read the will's boilerplate introduction and then came to a section directed specifically to Alex.

"It is my wish that my son, Alex, come under the care, legal custody, tutelage, and guidance of Professor Henri August Noblét at the French Academy of Sciences in Paris until he reaches the age of 21. Over the past 10 years, I have transferred funds to accommodate the development of my son's education, providing travel and a budget for his first five years under Noblét. This agent will provide the introductions and process my request. As for Noblét, he will know what to do."

Alex looked to his mother for confirmation, but her head remained bowed, as if she had heard nothing at all.

The agent continued reading Christian's last wishes. "I ask that my wife sign the necessary documents to allow transfer of custody. It is in the best interests of our child. I regret that circumstances as such make it impossible for me to provide anything more for my wife. She will become a ward of the state. We can only hope for a better future for our son in a different country and under a different system."

Alex snapped his glance up to the speaker, as if to protest. He could not imagine leaving his mother like this. Of course, there was no arguing. A will was a contract to be obeyed as closely as any law.

The voice continued, droning Christian's message to his son, "Alex, you can no longer remain in the U.K. Your future requires that you leave the country and go to a friendlier place that will make it possible for you to have a sustainable and valued future. Please accept this request as my last will and testament. One final thing: let me remind you what we learned together from our friend Carl Hiaasen: 'Good satire comes from anger. It comes from a sense of injustice, that there are wrongs in the world that need to be fixed. And what better place to get that well of venom and outrage boiling.' It is time for you to learn how to take the anger inside you and turn it to your own advantage. Alex, you are and will always remain the best of what I have been on this earth."

CHAPTER EIGHT

GALVIN LA CHAPELLE, LONDON

Mikhail finished his day working with *Manchester,* summarizing the project goals from the team meeting with Tharra.

At seven, he left the building and caught an autonomous taxi to the local gym to begin his workout. *Andréas* now took over. Having communicated with *Manchester* to understand Mikhail's state of mind at work, *Andréas* had already prepared a stress-reducing routine to transition Mikhail from work mode to a relaxing evening with friends. After confirming that this was, indeed, a social dinner, *Andréas* had "discussed" the dinner fare with *Georges,* the maître d' AI agent of Galvin La Chapelle restaurant, who had learned Mikhail's taste in French cuisine from *The Voice.* In less than a millisecond, Mikhail's evening had been prepared.

At the gym, Mikhail changed into his workout clothes, adjusted his earpiece, and walked out onto the gym floor. There he noticed another person at work on the rings. As he came closer, he recognized someone he thought was a member of the British Olympics gymnastics team. This was curious. Team members usually kept to themselves and never practiced away from the team gym, especially not in public.

Mikhail watched the athlete perform an extremely difficult maneuver with Olympic-class execution. Doing so, he became mesmerized with the motion of the figure as it moved through space. To execute with such precision, this routine required instinct, a well-honed mental and physical discipline, and great artistry.

Suddenly Mikhail remembered himself and looked away. Watching another person for so long was a social taboo. As he walked to the far corner of the gym to stretch, he felt embarrassed—and surprised—to have slipped like that.

Today's routine for Mikhail focused on stretching his muscular system. A few days earlier, he had completed a heavy weight-lifting session, and *Andréas* had noticed some breakdown of muscle tissue as a result of the excessive strain.

As *Andréas* coached him through his exercises, Mikhail's body responded but, unbeknown to *Andréas,* his mind returned to his project at work. Agents could read bodily functions but could not yet read minds. Mikhail's personal thoughts were still his own—for now.

Mikhail fantasized about the potential reach of his work. If he were able to create this sentient being, it would revolutionize an entire industry, yes, but it would also have ramifications well beyond any financial reward. On the other hand, he was playing with a new kind of high explosive; if he wasn't careful, it could detonate in the face of every living human being.

Mikhail hadn't been completely transparent with his colleagues earlier in the day. He hadn't shared that the initial tests performed on the *M* CNN tool set had proven faulty. So far, the human managers of CNN had failed to develop a tool that could identify potential errors or aberrant behaviors embedded in a sentient being. This was akin to having a mutant gene waiting to express itself into some terrible disease once unleashed in a body. It could take years to discover the mutation and understand the full consequences.

Mikhail had come to the conclusion that the world once again had lost its mind. How could anyone consider unleashing technology without thinking through the unintended consequences? In a way, he wasn't surprised. Scientists often pushed through the barriers of technology, creating advances without understanding whether humans were truly ready to manage all potential outcomes. Why should he feel any sense of responsibility to future generations, if clearly these scientists had not? His work would secure his status at the top of the social pyramid. He would be remembered as one of the greatest engineering architects. The real joy in this particular project was not in saving humankind from its own folly but in playing Master of the Universe—resetting the bar and changing entire industries. Although Mikhail would never admit it, he longed to earn the admiration of a brilliant scientist like Tharra in the process. He could imagine Tharra enumerating his successes in front of Nigel and the PAAI team. He would be the most famous man in the room—no, the most famous man in the *world.* Forever. No one could take that distinction away. His position at the pinnacle of society would be unassailable.

Mikhail contorted his body into a final yoga pose, elongating his neck, legs, and torso in the process, sweat soaking his tank top. He arose from this pose, stretched his legs one last time, and ran at full speed, executing a double somersault, landing perfectly on the gym floor.

Released from the task of controlling this exertion, his thoughts returned to the conversation earlier in the day: Would sentient beings have souls? How would these new creatures, with their ability to review millions of years of human behavior, decision making, and outcomes in a few milliseconds, impact humankind? Drawing conclusions without the benefit of ethics, empathy, or any of the basic elements that keep humans from killing one another—would they ever listen to mankind again? He knew the answer: a sentient being would never simply accept human domination or direction when it could outthink a human at machine speeds.

Mikhail caught himself before his mind dove further into this rabbit-hole thinking. He headed over to the rings, where he shook out his entire body, freeing his brain. He discarded his tank top, mounted the rings, and began his next routine.

Don't go there, he heard an inner voice say. *Those are dark thoughts. It doesn't help to think about it.* He tried to focus his mind on the present technical problem: executing the perfect iron cross. Suddenly, a vision of his boyhood friend Sasha flashed in his mind. The two boys were in the gym together, working through their high-bar routines. Sasha. What would he think about all of this?

Mikhail swung his body into his dismount, landing on the mat with a less-than-perfect bobble in his posture. Before he could berate himself, he noticed in his peripheral vision the man he'd watched earlier walking his way.

"Hello," said the figure.

Mikhail did a double take, shocked by the interruption.

"Don't be afraid. I just want to talk to you for a few minutes." The man was as tall as Mikhail. Barefoot and shirtless, he was the image of an ideal gymnast. He seemed to be European, but Mikhail couldn't identify his nationality.

Mikhail lifted his hand, signaling the other man to halt in place.

He ignored Mikhail's warning. "Do you mind if I share an observation?" he asked.

This was totally unacceptable, engaging a stranger in a conversation. Distance and aloofness—these were the proper behaviors.

"I noticed that you have a Russian style of execution," said the man. "Am I right to assume that your coach was a Russian?"

Andréas interrupted the conversation, informing Mikhail this stranger's facial scan indicated he was not a member of the British Olympic team but a French citizen visiting London. Without informing his master, *Andréas* sent a clandestine report of the incident to both *Manchester* and *The Voice.*

"Don't worry," the man said, "I don't take my agents into the gym with me. It destroys my self-awareness to have a device constantly correcting my every move." Mikhail noticed the man's hands were still covered with chalk. "I just thought I'd tell you I can help you improve your form with a simple change."

Mikhail nodded silently.

A familiar feeling arose in Mikhail. The man's language, the words he used, he sounded so very much like Sasha. A moment came back to him—at his childhood gym, Sasha crossing the room to offer Mikhail unsolicited advice.

The man took the rings from Mikhail, his hand brushing Mikhail's as Mikhail released the rings to this stranger. The stranger demonstrated a new way to execute the final move in Mikhail's routine. "See?" he asked. "When you relax your shoulder at just the right time. Right

. . . now—" the man launched, flipped through the air, and his two feet landed on the mat like posts driven into the ground.

Andréas interceded. "That technique is not recommended. There is no supporting evidence in the coaching manual that it works. Mikhail, I'd advise against using it." Mikhail ignored the recommendation.

As if he had heard *Andréas*' admonition, the man smirked. "Again, I never bring my earpiece to the gym. I don't like to be interrupted by agents," he said.

Mikhail took the rings, made the recommended adjustment, and landed in perfect form on the mat, just as the stranger had predicted. Mikhail grinned and turned toward the gymnast.

But as quickly as the man had engaged, he now had turned, picked up his shirt, and was walking across the gym floor.

Mikhail broke from his reticence. "Thank you," he called, as the stranger headed out of the gym.

Stopping in the doorway to the locker room, the stranger turned and called back to Mikhail, "My name is Alex. Remember your form. Perhaps I will see you here again. I work out at the end of every day."

Mikhail noticed that Alex had picked up an object as he walked into the locker room—a paperback. The government had banned all paper and hardcopy books with the advent of intelligent agents. Why did Alex have one in plain sight at the gym?

Mikhail shrugged it off. A gymnast who didn't use an agent to improve his routine. It seemed crazy. Yet Alex's recommendation had made all the difference. Apparently *Andréas* hadn't caught the problem with his form. How was that possible? Or was *Andréas* discouraging Mikhail from speaking with this stranger? He considered these possibilities as he ran the indoor track, logging six miles.

***

Mikhail decided to walk from the gym to the restaurant for his evening dinner engagement. It was another beautiful evening in London. Deep blue sky and the flowery scent of approaching summer days. The air was still this time of night, the temperature mild. The streets were filled with autonomous vehicles transporting people to their destinations. Mikhail felt a sense of exhilaration as he entered the old city, walking past the Roman wall and into an area that used to be covered with fields and nursery gardens. Known for its medieval history, this area was settled by the French Huguenots, silk merchants.

Galvin La Chapelle was a former red-brick Victorian-era girls' school with stone archways and an enormous open ceiling rising 30 meters above the floor. Mikhail remembered his LT manager taking him to this fancy restaurant when he first visited London. As he entered the building, his senses had been overwhelmed by the scent of flowers and spices. His manager had told him to order anything he wanted from the menu, and he had selected an extravagantly expensive bottle of French wine. That seemed like such a long time ago.

After a brief check-in with *Manchester*, Mikhail arrived at Galvin La Chapelle invigorated. The Michelin-rated restaurant was famous for its exceptional English and French cuisine, a special treat for a wonderful evening repast. Over the objections of the U.K. government, the restaurant did not permit outside AI agents—no exceptions—and for good reason. Years before, at a famous New York restaurant, a competitor had brought in an outside agent that analyzed the chef's award-winning recipes merely by accompanying the human host on a seven-course meal. The agent then multiplied and installed itself into the AI agents of the competing restaurant's staff. Galvin La Chapelle would take no such risks.

The decision to ban outside earpieces ultimately paid off handsomely, so much so it had been imitated by every great restaurant in the world. Galvin La Chapelle was now constantly sold out months in advance. It seemed that people wanted to enjoy a first-rate meal free of agent interference, interpretation, or outside communication.

Mikhail found Charles and Inès waiting for him at a table in the restaurant's main court. Inès wore the latest jeans and a flamboyant top; Charles looked tan and fit in his neatly pressed khakis and sweater. Both smiled so easily, such a contrast to the stone-like faces Mikhail saw every day.

Generally, Mikhail lived well within his means, but he savored an occasional meal in one of London's finest restaurants. He had high expectations for the evening. Regardless of what might transpire beyond dinner, he enjoyed the company of both Charles and Inès.

The couple shared a common interest in cinema, especially films from French directors such as François Truffaut. They also enjoyed a unique personal chemistry, finishing each other's sentences without mistake. When Mikhail first met them at a French film festival years ago, he mistook them for brother and sister. He soon realized they were close friends, sometime lovers, and nearly inseparable. Both were obsessed with cinema history and would quiz Mikhail with lines from old films to see how much he knew.

Charles had both charm and classic French good looks. "He looks like Gérard Blain in one of his short films," Inès once teased.

Charles retorted, "And you look like Ingrid Bergman."

It wasn't surprising that they had first met at a university festival celebrating movies from the last century.

Charles was a rugby player and blue-ocean sailor from the Noblét family. He wore the family crest on a small gold signet ring. His brother was a famous movie director who had won the Cannes Film Festival Palm d'Or Award. Still in his 20s, Charles himself was a noted psychology expert in the field of human interaction with machine intelligence. Inès, a bit older than Charles, had studied for her Ph.D. in French culture at the Sorbonne in Paris.

Neither Charles nor Inès appeared to have brought their earpieces to the restaurant with them. Extraordinary. Mikhail had dropped his into the pocket of his sport coat. Though he ordered his agents to turn off in compliance with the restaurant's policy, he knew earpieces didn't really have an "off" button. His agents could still monitor his activities or use nearby buildings to patch into the restaurant's network.

As Mikhail approached the table, Charles rose to greet him. He kissed Mikhail in the traditional French manner, before Mikhail circled the table to offer the same greeting to Inès.

"Mikhail. Lovely to see you," Inès said.

"You look so relaxed and fit," Charles added.

They settled into their richly upholstered seats. The table before them was laden with linen napkins, elegant place settings, candles casting a golden hue that played on the surface of the rosé already poured into crystal glasses. Around the room, black-tied waiters delivered food on silver plates as patrons sipped glasses of French champagne and European wine. Mikhail took it all in and then refocused his attention on his friends.

"Great to see both of you. Tell me the latest from Paris." Mikhail looked at Charles, but Inès responded first.

"We've just returned from a vacation in Cannes. We went to the Maeght Foundation in Saint-Paul de Vence. We wanted to see some works of Alberto Giacometti. Amazing sculpture. Do you know of him?"

"Of course. I have a piece of his displayed in my apartment. One of my favorite artists," Mikhail answered. "What did you like about his work?"

Charles interrupted Inès. "Everything at this exhibit is reduced to a feeling about the human core. His work expresses self-consciousness and frailty. His sculptures of human stick figures emulate the emotional states of the human condition. It's as though you can see into the soul of the individual when you experience his work."

Inès joined in. "It is an important topic these days, given where you are going with your work in artificial intelligence, Mikhail. Wouldn't you agree?"

"I really don't see the point," Mikhail said, guarding his true concerns. "Yes, there are decisions to be made about what we do with AI consciousness, but we have always had these kinds of decisions to make about technology. This is nothing new." He took a sip of the Domaine Tempier Bandol Rosé before him. No need to look at the menu—*Georges* had already selected this evening's courses.

"'Death is the mother of beauty,'" Charles said, quoting Wallace Stevens. "Do you think we can create a brilliant AI Picasso or AI Giacometti without the agent experiencing human emotion? How would it do this without the concept of mortality? Agents can learn all they want about how people feel, but it is another thing to actually experience the *feelings* of humanity. How does an AI agent understand what it means to experience the blue light of the sky in Cannes or the ancient beauty of the town of Saint-Paul de Vence?"

Inès added, "How would they translate what they experience into art that stirs the human heart?"

Mikhail continued to savor his rosé, steeling his face against his emotions. How had they plunged so quickly into such a discomfiting topic?

"Is that what your CEO Nigel wants to do? Allow people to use machine intelligence for financial gain alone, without any thought about the social consequences? Or perhaps he's in favor of using the power of artificial intelligence to create new masterpieces? In health care, managing a city, or what? Fine art?"

The questions annoyed Mikhail to no end. Yes, he had wondered about Nigel's intentions privately, but he didn't yet know where he settled on the issue, and he didn't particularly want to think about it now. He straightened up in his chair and looked directly into Charles's eyes. "Charles, this is boring. Let's change the subject. How did you find the Cannes Film Festival? Congratulations to your brother, by the way. Anything you would recommend to me? How is work?"

"Let me answer the last question first," Charles said. "Things are going well. We are making headway with this new focus on human and machine intelligence psychology. We are using personal history to create role-plays where patients can relive experiences from the past and engage in a way that changes their views on past outcomes."

Mikhail thought about his recurring nightmares. Nothing had helped him keep them at bay. "Is anyone practicing this locally?" he asked.

Charles raised his eyebrows and stole a glance at Inès. "Yes, actually, there's an amazing therapist practicing in London. Very skilled in this technique. He uses AI to create role modeling for adults to gain insight into early experiences."

Mikhail reached out to touch Charles's hand. "I am still having the nightmares."

Charles nodded and reached into the breast pocket of his sweater, retrieving a business card and pen. "I'll write down his contact information," he said.

Mikhail chuckled at the antiquated maneuver. "Who carries a pen anymore?"

Charles ignored the comment and slid the card across the table. Mikhail touched his hand once again, squeezed it for a moment. "Thank you," he said, taking the card.

The dinner conversation lightened from there, pivoting away from work and toward the libations at hand. The trio enjoyed two bottles of the Bandol, and a Margeaux recommended by *Georges*, to complement sumptuous courses of lasagne of crab with beurre nantais and pea shoots, followed by a main course of roast chateaubriand of Cumbrian beef.

Despite the pleasant company, Mikhail remained bothered by the outset of their conversation—an emotional reaction he couldn't quite name or shake. As they finished their dessert, they decided to retreat to Charles's nearby vacation rental to listen to music and enjoy after-dinner cordials.

How Charles had managed to find the only French-style apartment in all of London, Mikhail had no idea, but he appreciated the Louis Quinze furniture, French Impressionist paintings, and especially the fabulous cocktail bar. By the end of the evening, Mikhail felt a rare sense of freedom, having escaped his agents long enough to enjoy unguided conversation with real humans.

But there was also real danger here. The longer he stayed, the more Mikhail felt a persistent stirring, feelings long dormant from his youth. As that cloud funneled, he lost interest in the sexual encounter he had hoped for earlier in the day. Worse, he wasn't sure why his desire had flatlined, only that he suddenly felt compelled to get back to work. Perhaps he was just more excited about the work project? So instead of availing himself of the company, he apologized and made his exit. Though visibly disappointed, his companions handled his departure with characteristic grace.

"I really enjoyed this," Mikhail told them. "I am in Paris in a month. Let's the three of us catch a train to Cannes and spend a week in the city." While no one else in U.K. society enjoyed the option of travel to France, High Value Citizens could, as long as they delivered on their work contracts.

"Lovely," said Charles. "Let me see you off." Mikhail kissed Inès as she remained seated on the couch.

"Adieu, *mon amie*."

Charles escorted Mikhail to the door, where they paused. Charles placed his hand intimately on Mikhail's chest. Mikhail trailed his fingers down Charles's abdomen, then kissed him before disappearing out the door.

***

At home, Mikhail poured himself another glass of wine, removed his evening clothes, pulled on a pair of gym shorts, and sat down shirtless to resume his private project. Drunkenness didn't seem to deter his brilliance. But then, he didn't operate like a normal human being.

He began a series of tests on his algorithm design for a machine personality, creating a new architecture that would address each of the five *skandhas_Alexander* had mentioned in their philosophical conversation: matter, sensation, perception, mental formations, and lastly, consciousness. Each would have a machine equivalent in his model.

Mikhail had the schematic on how to create a Lloyd's Taiping dedicated quantum computing engine always accessible in the cloud. His own version of a sentient being would draw human values to direct behavior from assembled libraries and highly structured records of every category of information available. He would create the equivalent of a human brain, including the ability of independent decision making. Of course, it was not exactly what Tharra expected or wanted. Her vision was to create a powerful slave that would have a sole purpose: a perfect prediction machine Nigel could use to corner the financial and insurance markets. She had underestimated Mikhail and his willingness to chart a separate course.

For sensing the environment, the new form of consciousness could draw on a wide array of sensors, including imaging, sound, and touch. Robots were already capable of feeling an object and creating context for a measured, independent response. Mikhail would incorporate all this in his own designs.

Unlimited memory combined with super-fast machine learning would enable the new consciousness to draw judgments that could be checked against millions of historical instances. Mikhail already had unlimited access to storage and retrieval, so he merely had to design a way to leverage existing capabilities.

As for self-consciousness, Mikhail had to work on this last piece of architecture, to decide how to impact values through the *M* company's CNN architecture. *M* would allow him to load personality values into the architecture, making machine self-consciousness possible. But this was groundbreaking work—not well tested. He would need to design his backdoor code to be hidden in a critical last module, poised and ready to override any impulse on the part of his sentient being to do something he judged dangerous to human interests.

This would be the beginning of the work on the LT fail-safe switch. Tharra would never know that he went beyond a simple shut-off switch to actually making it possible to alter the LT personality. The key question in his mind: Should it be the LT personality Tharra wanted, or a personality that someone else had in mind? This was where his private project would cross over with his work at LT.

In Mikhail's mind, if the sentient being believes it was created by God, would it ever refuse a command it felt came from that God? Humans have the religious incentive of an afterlife. Machines don't die. So, what would motivate them? Would a machine require an ultimate authority to explain its own existence?

As the evening flowed into early morning, Mikhail finished his work for LT. But he wasn't yet ready for sleep. He had a personal project to engage in. He wanted to find out more about Alex, the stranger from the gym. Who was he? Where did he really come from? Was he as good a gymnast as he appeared to be? What about that book he was carrying? Mikhail instructed his agents to hunt for as much information as they could find on the mystery man.

Finally, before retiring for the night, Mikhail retrieved the card Charles had given him. He decided he would go through with a meeting. He sent a message to the therapist. Then he left his office for the night, leaving behind his wine glass for a kitchen robot to retrieve.

In his dressing area, Mikhail took off his shorts, dropping them on the floor beside his dinner clothes, where the laundry robots would take care of them. Naked, he crawled into bed and let *The Voice* manage the rest of the logistics—turning down the lights, setting the climate control, and assuming the care of its human charge.

As he moved into a deep sleep, Mikhail dreamed he saw an image of a Giacometti sculpture—a man's contorted body affixed to a plane of black ice that stretched to the horizon. In the sculpture's tortured expression, Mikhail recognized his own face, frozen in horror, staring down at a hole in the ice, a life-or-death decision to make. The nightmare had returned.

CHAPTER NINE

NOBLÉT

Henri August Noblét had enjoyed a long association with Christian Blake. It had started when Henri was a student at Imperial College's Engineering School. They both had studied artificial intelligence and had a common interest in French philosophy. They spent long hours together discussing Christian's theories, ranging from social contracts to the role of machine intelligence in human life. Henri had completed his Ph.D. at Imperial with a dissertation titled *Machine Intelligence: Inculcating Social Values*. It was, to say the least, a controversial topic.

Henri had come from a family whose history was defined by independent thinking with a focus on social and ethical values. His lineage derived from one of the few remaining Huguenot families in his ancestral home, Calais. They had been persecuted in the early days of the Reformation but had somehow learned to survive in a Catholic France. The original members of the family had been watchmakers, precision instruments that engendered no controversy. The dominant branch of the family became scientists and engineers, including Henri's father who was a professor of engineering in Paris before he died. Henri continued his father's legacy, focusing on artificial intelligence and machine learning.

"I've always stood at the intersection of human social values and machine learning," he had told Christian the last time they'd met, at a machine-learning conference in France, before Christian's fall from grace. "We are at a turning point. We must decide how we can embed human social values as a legitimate behavior algorithm for our machines, or we will ultimately lose our humanity as well as control over our destiny."

Christian thought there was merit to his friend's ideas but also thought Henri was taking the extreme view. Christian was an inveterate optimist at that time. Henri saw himself as a realist.

As engineers began to design machine intelligence to manage human behavior in the U.S. and U.K., France took a very different course. It kept machines subservient to human social values. The French were suspicious of private enterprise and the unfettered forces of capitalism. They feared that elites inevitably would use machines to produce even more wealth for themselves by controlling workers and the institutions that governed them. So the French took regulatory action early, resulting in the development of a very different machine intelligence strategy.

As Henri had observed: "In France, we saw the complete deterioration of the values of the American Republic to a point of no return. The events in America were a dagger pointed at the very heart of every democracy in Europe. Once our rights to privacy and control over information were lost, we would also lose the power to control what machine intelligence could do in the wrong hands." Henri saw this as the tipping point, the impetus for writing his controversial dissertation on machine intelligence. He wanted to stand up for the rights of humankind and make a bold statement to colleagues who might influence political direction.

***

A month after the reading of his father's will, Alex Blake found himself on a flight to France. Henri greeted Alex at the terminal. Alex had met Henri on several previous occasions. He'd even spent some time with Henri's son, Charles, while on vacation in the south of France when Christian was still a professor at Imperial. His father and Henri had so much in common—both of them having followed in the footsteps of their fathers, both sharing the same interests intellectually and professionally.

"So wonderful to see you again," Henri said, welcoming Alex with a huge smile and the traditional *la bise.* Alex reciprocated, albeit awkwardly. He wasn't used to expressions of emotion.

"So sorry to hear about your father," Henri looked Alex straight in the eye to express his heartfelt sympathy. "He was an inspiration to me and so many other colleagues. He will be greatly missed by all, especially those who loved his sense of integrity."

Alex didn't know what to say. Seeing his discomfort, Henri steered him to a private corner of the terminal.

"I didn't have the chance to even say goodbye," Alex said, awkwardly. "His last wish was that I come to France and finish my education with you. He had nothing but enormous respect for you. It had been a difficult few years for my father, as you know. Things are a lot worse for us in England." He diverted his glance to the floor. "I just wish I had had more time with him."

Henri nodded and lifted Alex's chin with his hand. "Let's get your things. I've arranged for a car to take us to Paris. You will feel more comfortable when we get you home."

***

On the drive to Henri's house, Alex peered out the window, awed by the freedom. Here people could drive into a city without fanfare or economic zone passes. He saw cars of every description, model year, and color—self-driving, autonomous, and old-fashioned manual drive.

Professor Henri smiled, "I suppose it must also be unusual for you to enter France with all your AI agents suspended at the border. We are not perfect. We still participate in the global bidding system. But we protect people's privacy and we maintain some civility over how we deal with our citizens. Here people still look each other in the face. They put away their devices when they engage in conversation. They are courteous with each other. Our agents are limited in what they can do."

Alex had noticed the differences. He found it disconcerting, almost invasive, when strangers looked him in the eye. He wondered if he'd ever get used to it.

"Once we get home," Henri said, "you will have an opportunity to get some sleep. Tomorrow, we can discuss next steps. Your father sent me a letter stating his intentions, which I will share with you. In your world, you are not used to making your own decisions, but my wife and I will want you to understand what lies ahead of you in the next few years."

As they entered Paris, Alex was amazed at the chaos. Traffic clogged the streets. Such a rich diversity of people walked freely along the sidewalks, unencumbered by police checks and zone passes. In the U.K., the press had always maintained that France was a renegade nation suffering from extreme poverty and deprivation. But that is not what Alex saw.

Maybe there really was an alternative to the way his own country managed its citizens' lives? The car passed by small shops, theaters, and street vendors—things he never experienced in London, and certainly not in Spitalfields. This was truly a sight to behold.

People were free to make their own decisions and control their personal information? What was that like? Such privacies had been eliminated long ago in the U.K. Fortunately for Alex, the U.K. government didn't care one way or another that the son of a *Zeroid* had left the country, as long as the U.K. received the required £100,000 exit fee. As Alex learned from the Family Services Office, his father had shown great foresight long before his demise, setting aside a percentage of his income each month, an insurance policy of sorts for his son's well-being.

As the car arrived in the driveway of a modernist glass townhome, Henri said: "Here we are! My wife has had a small meal prepared for us. I imagine you'll want a bite to eat before you sleep. Charles will be joining us tomorrow. I know the two of you will have a lot to catch up on."

"Arrivés," Alex said, jokingly. Professor Henri gave him a scowling look. His joke fell on deaf ears.

"In our country, you have the right to *vie privée*. Never forget that. Tomorrow we will deal with your visa and start the process to get you French citizenship. Your father had plans for your future. We're going to make sure they come to fruition."

Entering the townhouse, Alex felt a bit bewildered, though he had learned a long time ago not to express his feelings openly. How would he ever manage his life without agents? He couldn't imagine it. Under what kind of social contract would he be living?

Although he didn't let on, Alex had a thousand questions about it all. How would he function without a French-language agent? Did they expect him to learn the language through casual conversations? What about his mother, now living alone, a ward of the state in the U.K.? How could he accept this extravagant opportunity when his mother suffered so?

Alex didn't yet know that he would never see his mother again. She would fade into the *Zeroid* system. His father had enough resources to save only one of them. He had chosen his son over his wife.

LIFE IN PARIS

Alex's life with Professor Noblét proved transformative. Here was an alternative social order that could provide something he never experienced in England: privacy and real freedom. It was truly strange not to be constantly monitored or under the daily supervision of agents. Alex had become so dependent on his agents to make decisions for him that this new world overwhelmed him with its vast number of choices.

This wasn't his first time dealing with a major change. Moving to Spitalfields had been a huge transition for a boy of 10. From crisp personal servants to bored social service agents assigned by the government—everything had changed. The only things that remained the same were his father's tutorials, and those had kept his hope alive. He had matured rapidly through the adjustment and had grown more serious than most boys his age—even those in Spitalfields. Prohibited from visiting the city, he had missed London.

Now, from the darkness of Spitalfields to the City of Lights? His curiosity could not be contained. During his first year in Paris, at a dinner with Henri, Alex was asked to choose the wine. With no agent to choose for him, he froze. He had to ask Henri for help. "It's all right," Henri comforted him. "You've never been taught how to choose an appropriate wine or even how to ask the right questions to drive your decision. That's why you are here. I have a present for you," the professor said, pulling a small package out of his coat pocket. "It's a book on how to select wines."

Henri's son, Charles, loved gymnastics and was already extremely proficient in the sport. He thought this might be a way to reach Alex, so he encouraged the boy to take lessons. At first, Alex felt awkward doing gymnastics, but he soon learned that he enjoyed the physical challenge, and he could use it as a way to focus his mind. Onlookers responded well to his routines. Appreciating the attention, he trained even harder.

Dinner conversations between Alex and the professor explored many topics, but frequently they turned to the economic systems of the Western countries. "I can't comprehend the French system," Alex once told Henri. "How can anyone challenge the efficiency of the U.K. or U.S. systems?"

The professor had calmly replied, "You'll see. People in Britain and America can't understand the value of personal choice because the vast majority have never experienced it. From the day they were born, decisions were made for them by machines and omnipotent systems. So they don't know what it is to be truly free and human. Worse, they don't know what they don't know. Give yourself some time, Alex," Henri said. "It's confusing now, but you'll find that individual liberty is the single most important thing a human can experience. You'll be thankful you discovered it before it was too late."

Over the course of the next few years, Alex adapted to his new country so well that an outside observer would never have guessed he had spent his early years in England. He became fluent in French, learned to love gymnastics as a way to instill self-discipline, and acquired a brother, Charles, who treated him as though he had been born into the family.

ALEX AND CHARLES

Charles Noblét was Alex's age. The two of them had become fast friends when their families had vacationed together in the south of France. They had hiked the limestone valleys leading to the Côte d'Azur. Through long evenings they eavesdropped on the adults' dinner conversations, and they fell asleep together playing word games in front of the fireplace. It was a wonderful time. As Alex made the adjustment to life in France, Charles had become his principal emotional support.

In contrast to Alex's serious demeanor, Charles lit up the world with his energy and sheer optimism. He loved everything about his life, and he drove Alex crazy with his endless flow of new ideas, wild childhood stories, and zest for living. He was also an ardent lover of cinema. Charles compared all of life's events to scenes from classic movies. Indeed, famous movie scenes became tropes in his everyday conversations: "Snakes! I hate snakes." "Rosebud." "I'm walking here!" "You can't handle the truth" (which he often said to Alex), "Well, nobody's perfect," and a thousand more. The two friends played games, firing off movie lines and trying to outguess each other, identifying the movies they came from. Charles also enjoyed sports and coached Alex in gymnastics.

Alex and Charles spent many days and nights together, debating every topic under the sun. Alex began to learn how to have an opinion, think for himself, and defend his points of view. Professor Noblét raised his two sons to be independent thinkers and gave each of them an unusually wide berth to pursue his own interests. All ties to England had now been severed, and Alex became a fully engaged French citizen. He finally renounced his English citizenship in his mid-20s.

When the time came, the two young men decided to enter the Sorbonne together. Charles studied with a pioneering psychologist who explored the human behavior issues raised by technologic advancement and machine intelligence. Alex, meanwhile, studied political science and then decided to follow Charles into psychology. His fluency in French was now complemented by a thorough understanding of French philosophy, political discourse, and—of all things—human-machine intelligence.

Charles was now tall and athletic, with the face of a handsome aesthete. He was still sunny in his disposition, thoughtful, and now more soft-spoken. Alex was almost a head taller and darker but, thanks to the gymnastics, he was muscular and moved as gracefully as a ballet dancer.

At university, Charles met Inès. She was also interested in movies. When Alex met her for the first time, they bonded instantly. The three of them would form their own fan club for classic movies. Inès was more of an intellectual, and she also possessed a rebellious spirit. The combination of the three of them was a bit dangerous.

After graduation, Charles became a well-regarded therapist, practicing in Nice with an office at the Sophia Antipolis University. His work focused on trauma inflicted by disruptive technology. He would later introduce Alex to a famous French therapist in Nice who was planning to establish his practice in London, with the goal of working with patients adjusting to becoming High Value Citizens. The three of them met at a local bar during one of Alex's frequent trips to Nice.

In their final year at university, Charles, Inès and Professor Noblét had become involved in a new French underground movement, *Intelligence artificielle pour l'âme de l'humanité* or simply *L'âme.* Alex knew of the group's creation, but oddly, he wasn't invited to join. Hurt, he approached Henri, who told him: "This isn't for you right now. Perhaps later. Indeed, it might prove to be a handicap for you to be involved. Just be patient: there may be other ways for you to contribute."

L'âme formed secret chapters throughout France, Germany, the U.K., and the U.S. Its mission: restore privacy as a right guaranteed by law, even if it meant sabotaging artificial intelligence platforms targeted at human society. Members included some of the most talented computer scientists, philosophers, and political scientists from around the world. Society members used secret cells in their network that were kept hidden from computer systems in order to remain undetected. Charles and Inès were assigned as agents to set up a cell in the U.K. with the purpose of infiltrating the engineering communities in that country.

Meanwhile, their regular lives were, to all appearances, perfectly normal for young people in their 20s—parties, the working world, endless philosophical discussions, love affairs. Alex was with Charles and Inès through all of this, and was their closest friend and intimate. But when he asked about *L'âme,* both would merely smile and change the subject. Alex would walk away upset that this was the one area he couldn't share with his friends. Finally, he decided he would bide his time but that he would ultimately find a way to support their cause. He, more than anyone else, understood what was at stake.

CHAPTER TEN

THE PLOT

A month before Mikhail sat down to dinner with Charles and Inès in London, Henri had asked Alex to join him for wine and cigars at his home. It was a beautiful June afternoon in Paris, 10 years after he had come to live with the professor, and two years after he left university. The professor opened a bottle of fine red wine and beckoned Alex to sit next to him.

"I know you've felt left in the dark for a while about *L'âme,*" Henri said, in his aristocratic French. "Well, the time has come to tell you the whole story."

Alex allowed a slow smile. "I was almost afraid you'd given up on me."

"Oh no, my boy," the professor replied. "In many ways, your contribution will be more important than any of ours, your destiny perhaps one of the greatest."

"I don't understand." Alex sniffed the wine, then took a small sip, rolling it around on his tongue.

Henri smiled knowingly. It was a fine Bordeaux, and now Alex could appreciate it. "Do you remember the French psychologist that you and Charles met years ago?"

"I do," Alex nodded. "He specialized in machine intelligence and those that had experienced trauma adapting to the new social order."

"Yes, that's him." Henri fished two cigars from his pocket and handed Alex a Cubano. "He is now practicing in London."

Alex lit both their cigars.

"He has since changed his focus. He now works with machine intelligence to re-create past traumatic events, capturing the attributes of real human behavior from past actors in people's lives, and then allowing people to interact and change outcomes. Really amazing work."

"That sounds unbelievable," Alex said, lifting his cigar to his lips.

"Patients can ask questions of figures from their past or probe their behavior to gain greater understanding of what really happened during traumatic events. And that helps them interpret their own response."

Alex exhaled cigar smoke. "Like a form of time travel. *Twilight Zone*."

"Indeed," Henri replied. "The AI agents can simulate a reenacted engagement, and the results are astounding, with very high cure rates for depression and other behavioral anomalies brought on by past trauma."

"Interesting stuff," said Alex. "But is that the reason you called me here today?"

"In part, yes," said Henri. "I want you to travel to London to visit this therapist and learn his techniques. I'll provide you with the necessary introductions and a place to stay. I know you may be apprehensive about going back to London and its way of life, but you will have to trust me when I tell you that this trip is of vital importance."

Alex nodded gravely, "Then I will do as you ask. If I may ask, is this part of the work you are doing with Charles and Inès?"

Henri put down both his cigar and wine, and leaned toward Alex. "What I am about to tell you must remain strictly confidential. You must never talk about this or any other conversation we have had about the therapist or the purpose of your visit."

Alex nodded.

"The London therapist is a member of *L'âme*. Your mission is simple. You will quietly approach and befriend one of his patients, a prominent AI scientist who is in the center of new work being done on advanced machine intelligence. The name of the patient is Mikhail.

You must subtly approach him, befriend him, and be there to listen to him as a friend helping him through his therapy. You must help him open up to his past feelings and become more of the person he once was."

Alex nodded, intent on playing his role well.

Henri continued, "You must never mention your past life, who you are now, or that you know Charles and Inès. It is vitally important."

Alex drained the final sip from his wine glass, thinking about this assignment. Going back to England? To that world? He had never imagined doing such a thing. After years of pestering Charles and Inès for information about *L'âme,* he desperately wanted to understand the group's work. Maybe now was his chance to find out? "What more can you tell me about *L'âme*? What is our goal here?"

"The world is on the verge of developing a sentient being. We believe that when quantum computing becomes available, the personality and value system of the new machine will make it impossible for any human to control this machine's decision making. This is a turning point for the human race. We all know what the U.S. and U.K. already have done with machine intelligence, regulating jobs, social stature, and quality of life for citizens in their society. *Zeroids* are only a step away from being slaves."

A picture of Alex's mother flashed in his mind. No one should have to live that way.

"Mikhail is poised to be instrumental in this process," Henri continued. "Your goal is to influence some of Mikhail's thinking, to help him awaken from a very long sleep that began in his childhood, to find his humanity again. You will understand more in due time. Initially your job is simply to become a trusted friend which, as you know, will be extremely difficult in that society these days. You will have your work cut out for you."

Alex took this in. Approach a stranger? In England? "I need to think through my strategy," Alex said. "What can you tell me about Mikhail?"

"Mikhail loves gymnastics, as do you. We believe he is searching for meaning or at least some sense of peace. He experienced the death of a close friend, whom he loved, when he was a child. This particular therapist will help him through that experience, help him open up what he shut down so many years ago. It could be transformative."

Alex looked at Henri, saw the gravity in his face, and knew how important it was to complete his mission.

"We also believe Mikhail is bisexual, so you can use that in your work with him. Really, all you have to do is be yourself around him, and I trust it will all work out."

Alex nodded.

"I cannot impress upon you enough how important this mission is. Mikhail will someday be in a position to protect or surrender all our liberties. You and the therapist will work independently of each other. While he runs the therapy sessions, we need you to become a key influence in Mikhail's life. Introduce him to new ideas that will encourage him to think about the consequences of his work—without specifically talking about that work."

"Does Charles know this man?" Alex asked.

"Yes. You will have a chance to learn more about him when Charles comes for a visit next week. We will ask you to leave for London after the visit." Henri arose from his chair and touched Alex's shoulder. "We will tell you where and when the subject works out, and it will be up to you to decide how to approach him. Are you willing?"

Alex nodded, pleased, finally, to be called on. "Of course."

***

Over the course of the previous decade, Alex had changed his mind about the U.K. and its "more efficient" system. He now understood how AI agents in the hire of his father's enemies had destroyed an honest man's career. He had learned what it meant to lose his and his family's liberty. He wasn't prepared to lose that liberty again at the bidding of some kind of master machine.

Henri had prepared a fake research project to give Alex an official reason to travel to England. When he landed at Heathrow, Alex could already feel the mental chains returning. Border police not only questioned him about having any ties with the French government but also admonished him over his visa, emphasizing restrictions, including the fact that he could travel within London only during limited hours.

Further, he was permitted to lodge in only one city outside London, Spitalfields, which was the subject of his study.

Over the next few days, Alex met with the therapist and began to work out at the designated gym. Charles and Inès had already arranged a dinner with Mikhail in London to plant the seed. Alex's mission was to nurture and grow it.

CHAPTER ELEVEN

LUBYANKA SQUARE, MOSCOW

THE RUSSIAN PLOT

The Solovetsky Stone is situated in Lubyanka Square, directly across from the Federal Security Service of the Russian Federation. It was dedicated on October 30, 1990, to honor the memories of the millions who died in forced-labor gulags during the era of the Soviet Union. Here, in this square, before this monument, Mikhail's former coach, Andréas, had scheduled an urgent meeting.

Andréas had taken the new high-speed train service from Saint Petersburg to Moscow, traveling 400 miles in under two hours. The train system was built as a joint venture among Siemens, Thyssen Krupp, and the Russian government, as a symbol of the recent Russian cooperation with the European Union.

As he arrived in Moscow, Andréas decided to walk to the square. As he approached the monument, he looked for a special earpiece, finding it exactly where he was told it would be: in a garden located near the monument. He activated it.

Long ago, the Russian government had created a "no signal zone" around this monument and other secret locations throughout Moscow to prevent anyone from listening in on conversations or intercepting messaging communications. In these spots, only earpieces activated for specific missions could receive and broadcast transmissions.

Andréas Vasili Sokolov was an agent of the Federal Security Service. He was trained directly out of the Russian Special Forces and placed in academic environments to befriend, recruit, and eventually train Russian nationals living abroad. Over the course of two decades, he had placed 20 Russian agents in 12 different countries.

With the advent of His Master's Voice a decade earlier, and its revolutionary voice-operating system that enabled agents to manage people's lives, a new opportunity had emerged. His Master's Voice machine intelligence created agents that were now omnipresent *and* could learn every minute detail about a person's life. Every AI agent was now connected to every network, everywhere in the world, all the time. Already, agents had been created with subtle personalities, designed to transfer emotional connections between human and AI agents, building trust.

All the work of understanding how to drive human behavior was harnessed in the cloud and delivered through the simple device of a headset: a miniature neuromorphic supercomputer connected to a vast system with enough power and storage to remember all of human history down to the tiniest detail. Now, the new platform needed only the brilliance of a Russian leader to intercede to direct the technology to serve Russian interests.

Vladimir Vladimirovich Putin, the architect of what had become known as the "Putin American Plan," had been that sort of insightful Russian leader. His government's earlier intrusion into the U.S. election system had discredited American democracy. The Russian government had employed cyberattacks and planted alternative facts in the news media and social networks, all with the aid and comfort of the American people and their foolish legislative representatives. Putin ruled Russia as president until he died in 2041, a national hero. His ultimate dream was to find that one opportunity to enter and control the network connecting all of humanity—in this case, penetrating the architecture of an intelligent, self-thinking entity that could be an integral component of His Master's Voice.

Putin's dream was becoming reality. His handpicked successor continued and expanded his program. With the founding of *M* and its invention of the Consciousness Neural Network, it was now possible to develop a deep Russian agent intelligence that could learn, make its own decisions, and be implanted in every machine to subvert Western democracies and institutions. This was the great leveling of the playing field: Russian cyberwarfare expertise in place of Western brute military force.

Two tasks remained: (1) create a Russian personality implant that reflected Russian values, government strategy, operating directives, and loyalty; and then (2) secretly make it part of every human AI agent system in the world, through the *M* tool. This sentient being would protect and defend Russian interests and would finance itself by taking a percentage of every equity financial transaction globally, then influencing AI agents to bend to its will. With *M*, the goals of the Russian government had aligned themselves with corporations seeking to use the technology to control their customers' purchasing behavior.

As Andréas turned on his earpiece, a familiar AI agent whispered in his ear: "Announce code to access."

"KP. Code *Manna*." Andréas named the access code KP after the famous Russian spy, Kim Philby. He used the name of a gymnastics move, *manna*, as the execution command to start a secret project designed to deliver Putin's grand vision.

"This is a localized broadcast and cannot be traced or recorded. M*anna* activated," the device replied.

"Ready to proceed to next steps," said Andréas.

"Proceed and report back progress two weeks from today—exact time and place," the device instructed.

"Acknowledged. Operations Plan Manna next steps proceeding to execution. Returning two weeks from today—exact time and place."

"Transmission ended."

Andréas returned the earpiece to its case, replaced it in the hidden location, and walked away from the square. Within 60 seconds, the earpiece would self-destruct. Operations Plan Manna would set in motion the activation of a deep Russian agent already designed into the architecture of the His Master's Voice network. Andréas would contact a key Russian human agent to complete the execution of the plan.

Project *Manna* would expand the global reach of its client by replicating itself in the Consciousness Neural Network's core code. With both His Master's Voice and CNN penetrated, it would be possible to run the first experiments testing the ability to control people's behaviors on a massive scale. As part of this plan, Russia would use its cyberwarfare expertise to siphon a billion U.S. dollars to fund the scheme.

CHAPTER TWELVE

MOSCOW'S WEST 4 CAFÉ AND THE BACKDOOR

A few weeks after her team meeting in London, Tharra Warrior arrived at the Moscow airport, accompanied by her father, who was to speak at a conference on Russian-Indian affairs.

They took the local transit to the Four Seasons Moscow Hotel. As she checked into her room, Tharra found a handwritten message on the entry-hall table. It was from an old friend, asking her to meet that evening at West 4, a favorite coffeehouse.

Hours later, as she entered the café, she saw Andréas, whom she had known since she was a child when he had served as a young Russian military adviser to the Indian government. Andréas had gotten to know Tharra's father well back then. What Tharra later found out was that Andréas had left the military years earlier to join the Russian intelligence services. His official cover was that of a university gymnastics coach. Mikhail was one of many students he had coached.

In fact, it was Andréas who had arranged for Mikhail to be hired at Lloyd's Taiping before Tharra had joined the company. Andréas had all the pieces in place for Russia to accomplish its mission. Now he had to work with Tharra as the last puzzle piece of the Russian plot.

Tharra hadn't seen Andréas in years. She noted how the scarring on his face had become deeply creviced with age. The two settled in at a corner table and ordered their usual drinks: a double espresso for Andréas, chai for Tharra.

After the obligatory preamble of catching up on each other's lives, Andréas broached the real subject as their drinks arrived: "We have a mutual acquaintance, Mikhail Ivanovich Vasiliev," Andréas said. "He was a student of mine when I coached gymnastics at his school. Fine gymnast and very bright boy. He works with you in London, yes?"

Tharra raised her eyebrows—this was a surprise. "Yes, he leads one of our teams. Very bright, or so I've been told." Tharra picked a sugar cube off the saucer under her teacup and dropped it into the tea, erasing any emotion from her face. Was Mikhail a Russian agent?

"Actually, that's an understatement," Andréas continued. "He is perhaps one of the most talented computer-science engineers in the world today, and he is working for you. You see, I know a lot about this young man, and not just about his accomplishments but also about who he is as a person. Valuable insights." He let that idea rest for a moment as he sipped his espresso, wincing at its temperature as the liquid touched his lips.

Tharra decided to cut to the chase. "Why are you telling me all this? What is your interest in him? Is he a Russian agent?"

"No. It is really about my interest in you," he said, placing his demitasse on the table. "When your father and I met many years ago, Russians aided your country just when it needed our help. When your father's own government turned away from him during their rapprochement with America, he came to Russia again, this time to work with us on our own projects."

"Both my father and I appreciate what you and the Russian government have done for us over the years," Tharra said, a note of caution in her voice.

"It is now time for us to talk about our need for your help," Andréas nodded. "It turns out your father did more than return the favor. He gave us certain intelligence on nuclear weapons locations. He wanted us to ensure that he would have a permanent home and position in

Russia in return for the information. He also wanted us to ensure your placement at MIT to work on the *M* project. We were quite willing to help with all of this."

Tharra stirred the now-dissolved sugar into her tea, uncertain where the conversation was going.

"The truth is, my dear, your father is a great friend of Russia, although we won't speak about the nature of that friendship just yet. A bit premature. You certainly would have gotten into the MIT Schwarzman College of Computing without our help, but it didn't hurt to have recommendations submitted from the head of the admissions team, who later would become one of the founders of *M*."

Andréas grimaced a bit as he prepared for the next part of the discussion. Stalling, he touched the side of his espresso cup, testing the temperature. "It is now time for me to give you some more information to consider." He lifted the cup and drained it in one long swallow.

"We want you to continue to pursue your efforts at Lloyd's Taiping with full confidence and vigor. What we ask you to do is to combine your efforts with Mikhail's to leverage the backdoor you intend to create into the CNN architecture, your so-called fail-safe switch."

Despite her best efforts to remain stoic, the surprise registered on Tharra's face.

Andréas noticed. "Yes, we know all about this. This backdoor in the code will allow us to insert special software modules from time to time that need to be undetectable." Andréas smiled. "A fitting challenge for your talents, wouldn't you say? Of course, Mikhail must not know about us or our true intent. So you must find a way to get him to deliver what we need."

Tharra's face grew grim. "This is all rather problematic. Why would I risk everything, including my own reputation, to do this for you?"

"Three reasons," Andréas replied. "First, it will ensure that the Indian government never finds out about your father's treason. Second, you and your father will be guaranteed protection in Russia once the mission is complete. Third, you will be paid $50 million up front in bitcoin for creating the backdoor, and another $50 million once we are assured that it works."

Andréas surveyed the room. The usual crowd of individuals and pairs were talking to their earpieces, oblivious to their surroundings, while Tharra absorbed this information. When he looked back at Tharra, her eyes had grown cold. Perhaps she would need some coaxing. "Consider the alternative scenario, my dear, if you decide not to help us out," he offered. "It's not a very pleasant one, I assure you."

Tharra maintained her steely gaze. Leaning forward, she delivered her next words with force: "Those are three compelling reasons, but it isn't necessary to threaten me or my father. This will require more money than you have outlined and certain assurances that we can get out of the U.K. in one piece. We can speak about that later."

Tharra straightened up, seemingly unfazed by Andréas' request, as if the whole show were in her control. "Now, how much do you know about Mikhail? How do you really know what motivates him?"

Andréas grinned in an attempt to appear relaxed, but translated through his facial disfigurement, he appeared more tense. "As I have said, I have known Mikhail since he was a child. He lost his emotional connection a long time ago, when he witnessed the accidental death of his best friend. Since that time, he has focused assiduously, to the exclusion of all else, on his technical work as a means to bury the memory of his friend's death. As a result, he has a brilliant mind, yes, but under all that he's vulnerable. With the right direction from you, he will play right into our hands."

Tharra looked off to the distance, considering, Andréas thought, how she could manipulate Mikhail's weak spots.

"Mikhail wants neither money, nor love, nor sex. He wants recognition as an engineering superstar, validation that his life's work and dedication have paid off, that he's reached the upper echelon even among the elite. With your understanding of *M* architectures and Mikhail's efforts to create a hidden backdoor, we will be in a position to insert our module into the architecture undetected."

Nonplussed, Tharra brought her attention back to the table, taking a sip of her tea.

Andréas continued, "Many years ago, the Russian government ordered bot installations on your and Mikhail's agents, along with that of your father's. This order came from the highest authority in our government. We have been monitoring Mikhail 24/7, including every major interaction, since he was seven years old, when his talent was first identified in Russia. Our predictive analytics team has war-gamed him multiple times and knows how he reacts to virtually every situation. He has never questioned authority, nor the English way of life. Nor is he self-aware, other than the work he does on the gymnastics floor, which never leaves the gym."

Andréas leaned forward. "We also know who you are and that you have known, for a long time, that you were also being monitored. Interestingly, you did nothing about it."

Tharra didn't react, simply set her teacup back on the saucer.

"In many ways, you and Mikhail are kindred souls. That said, we understand you are far more motivated by money than he is. If you do what we ask, you will step into the center of a very big storm that will define artificial intelligence—and the collective human story—from here on out. You will be as well-known as Oppenheimer and the atom bomb, and you will profit from it enormously. And Russia will once again become the dominant power in the world. And you will be here with us to enjoy it. It is our destiny."

Tharra finally smiled. "Indeed. We will talk again. For now, I will consider what you have said and do my own due diligence. I will call for a meeting, and you will have my answer."

"I expect a positive answer—soon," Andréas warned. "Or there will be consequences." He couldn't have been more direct.

Tharra stood up, gathered her things, and left the café. As she did, Andréas removed from his pocket a device that had blocked any recordings and scrambled any transmissions of the conversation, should Tharra have decided to capture the event. He rose confidently from his chair, left money for the bill, and walked out to catch the train back to Saint Petersburg.

CHAPTER THIRTEEN

SAINT PETERSBURG, LAKE LADOGA

Mikhail had grown up among the spectacular architecture and bucolic environs of Saint Petersburg, where his family had lived for five generations.

Mikhail lost his mother to cancer when he was four years old. His father, a kind man who appreciated art and culture, died a year later in a plane crash while on a business trip to China. Mikhail's aunt and uncle raised him as their only child.

With only faint memories of his father and no recollection of his mother, Mikhail grew up in the city with his aunt and uncle, art historians who worked at the Hermitage Museum. Every December 7, they would bring Mikhail to a private party for museum patrons to celebrate its founding.

There, Mikhail met a diverse array of people from every walk of life, including businesspeople, art patrons, artists, and musicians. It was an eye-opening experience for a young man. It was at one of these parties where a family friend introduced Mikhail and his uncle to a famous gymnastics coach, Andréas.

In the winters, Mikhail spent his late afternoons at the children's gymnastics club. There he learned the skills of self-discipline and control while also experiencing his first brush with athletic competition. Early on, his coaches believed he possessed an extraordinary talent for the sport.

In school, Mikhail was also recognized by his teachers for his skills in math and computer science. When he wasn't engaged in schoolwork (supervised by his uncle) or in gymnastics, Mikhail used his weekends to explore the forests and lakes around the city, learning how to ice skate and hunt.

Wherever he was, Mikhail displayed a wonderful smile that attracted people to him. He told funny stories and lit up every room he entered. During those years, the boy seemed to have limitless energy and infected others with his enthusiasm and optimism. One particular friend came along, amplifying the boy's happiness.

Mikhail's first encounter with Sasha occurred at the club. During a regular workout, Mikhail saw a group of kids gathered near the rings, watching a new gymnast who had recently moved to Saint Petersburg from a rural village some 30 miles away. As Mikhail approached the crowd, he noticed one of his coaches focused on the young performer, who had just executed a perfect set of double somersaults on the high bar.

Sasha was only 12 years old, the same age as Mikhail, but he had already mastered the bar like a true Olympian. Mikhail's first reaction? An uncharacteristic explosion of competitive jealousy. When it came to athletics or academics, his status had never been challenged by his peers, let alone by a newcomer. He didn't like it.

The other kids clearly saw in Sasha a new athletic role model, and they were open about their admiration—too open for Mikhail's tastes. It had always been Mikhail who had received that attention. Now there was a new king.

As Sasha dismounted the bar, Mikhail took stock of his competition. Sasha had long blonde hair, a well-built gymnastic torso, and a brilliant smile. Yet Mikhail detected something soft about him, a gentleness, an absence of any apparent arrogance or haughtiness, despite his brilliant performance.

As Sasha walked over to his coach to receive feedback, he glanced at Mikhail. For an instant, their eyes met. Then Sasha made his way to the sidelines for water and a towel.

There Mikhail approached Sasha with caution, introducing himself stiffly. "Hello. I am Mikhail Ivanovich Vasiliev, a student of Andréas. I watched you on the bar. Your work is better than anyone I have ever seen, except, of course, for myself. Congratulations." Mikhail reached out and offered Sasha his hand.

But he received no reaction. Not a handshake, not even a word. Sasha continued to wipe the sweat off his body with a towel.

Mikhail pressed on. "I don't know much about you, but my coach, Andréas, told me you recently moved back to Saint Petersburg?"

Sasha looked Mikhail directly in the eyes. "He is also *my* coach now. I had no former coach before him. I learned how to do my routine from online training videos. You?"

Mikhail couldn't believe what he had heard. "How is that possible?"

"It's possible. I need to go now. My uncle is picking me up to take me home. Perhaps we will meet again." Sasha grabbed his gear and headed toward the exit. An older man greeted him at the door and took him to a car waiting outside.

Eventually Mikhail would learn that Sasha's father had recently divorced his mother. She sent Sasha to her brother to raise Sasha in the short term as she tried to straighten out her life. Like Mikhail, Sasha had poured himself into sports to get through the stress. He loved both his parents, and they loved him, but his father had had an affair and it was just too much for his mother to endure—and so the divorce.

Despite his intense work ethic, Sasha appeared understated in the way he approached his life. He exceeded everyone's expectations at everything he did, whether it was his school studies or gymnastics, but he never exceeded his own expectations.

Thinking about that first encounter years later, Mikhail would remember how self-confident Sasha had seemed. When he talked to his coach about Sasha, Andréas had told Mikhail he would need to get used to having a competitor in his world, that the experience was good for him because now he would have to perfect his form and focus on keeping up with the competition. In Mikhail's mind, his coach seemed to assume Sasha had already taken leadership away from Mikhail. It was deeply humiliating.

There were more surprises to come.

Despite Sasha's initial diffidence, both boys were naturally drawn to each other because of their mutual love of the sport, and because of some kind of chemical reaction. They both were enthusiastic, motivated to accomplish great things in life, and passionate about their athletics.

Andréas had a different view of their relationship and goaded both boys to compete with each other while he taught each of them to "borrow" characteristics of the other to learn how to perform more effectively. He encouraged Sasha to emulate Mikhail's competitiveness and Mikhail to learn how to read human nature as well as Sasha did.

The Russian schools had adopted technology early on to educate and train talented students. AI agents were assigned to augment teachers in the school. The agents were designed to influence as well as educate their charges. China had implemented a similar program many years earlier than the Russians. All video devices were carefully controlled, and content was managed not just by parents but by the government and school systems.

Just as in the U.S. and the U.K., Russia and China understood the addictive power of video devices and wanted to harness that power for their own purposes. The Americans took it further, allowing preschool children to be raised by their devices, creating fantasy worlds to capture emotional connection early on that could be repurposed later in adult life. In the U.S., private technology companies were relentless in their pursuits of early-age "market capture," as they called it.

Under this system, Mikhail was doing advanced programming by the time he was eight years old. He had a natural affinity for solving complex problems working with machines, and he enjoyed it thoroughly. This did not escape Andréas' attention, since the Russian government's policy was to focus on AI as a weapon to counter the West.

In contrast, Sasha mastered literature and art, which was not of much use to the Russian government. No matter: Andréas' real interest was Mikhail. Sasha was simply a catalyst to keep Mikhail happy.

Like Mikhail's, Sasha's trajectory was established early. Alexander Sergeyevich Pushkin was Sasha's literary hero. At the age of 10, Sasha had read most of Pushkin's early works, including *Eugene Onegin.* He marveled at each character's radically different approach to life. As Mikhail's natural affinity toward mathematics transitioned into computer programming, Sasha became Mikhail's counterpart in literature and art. Sasha had already begun writing his own poems and short stories. He saw literature as a way to understand human nature, and this honed his skills in learning how to read people, a skill Mikhail sorely lacked and wanted to learn.

In short order, the two boys became inseparable friends. They spent nearly every waking hour with each other, in school, at the gym, and at each other's houses, discussing this or that topic or practicing gymnastics and spurring each other on. They explored Saint Petersburg together, a place they both regarded as Russia's most magnificent and beautiful vision, nestled on the Neva River.

When the boys turned 15, they spent more of their time skating on the lakes and rivers around the city and taking excursions to country dachas owned by family friends. Sasha's dream was to publish his first literary work by the end of that year, just as Pushkin had published his first work at age 15, while living in the Court of Peter the Great.

On Christmas day of Sasha's 15th year, the two boys were skating on the ice of Lake Ladoga. The sky was blue from the horizon, rising to a ceiling of cirrus clouds filled with ice crystals created by the Russian winter. Only a few skaters were out on the ice that day, most of them near the dam at the end of the shore. An observer from shore would later tell the police that he saw the two boys skating together when suddenly Sasha broke from Mikhail, racing in the opposite direction, toward the dam. It took Mikhail a while to realize his friend was no longer following him. The observer heard a fierce cry as Sasha broke through the ice. Several nearby skaters tried to rescue the boy, but all in vain as the frigid water took its toll.

Mikhail was devastated by the accident, nearly inconsolable. In the weeks and months that followed, he drew into himself and experienced the first bouts of severe depression that would define his mood for years to come.

Along with Mikhail's family, Andréas increased the time he spent with Mikhail to nurse him through the depression. But Mikhail found his own direction. He became obsessed with his programming, using it as a weapon to fight off his demons. He became aloof to his friends and teammates, and spent long evenings at home by himself, learning new approaches to algorithm development and artificial intelligence. Seeing a future unfold before his student, Andréas encouraged Mikhail's obsession.

CHAPTER FOURTEEN

THE LONDON THERAPIST

'Tis a fearful thing
To love what death can touch.

A fearful thing
To love, to hope, to dream, to be—

To be,
And oh, to lose.

A thing for fools, this,

And a holy thing,

A holy thing
to love.

For your life has lived in me,
Your laugh once lifted me,
Your word was gift to me.

To remember this brings painful joy.

'Tis a human thing, love,
A holy thing, to love
What death has touched.

—"'Tis a Fearful Thing," Yehuda HaLevi[2]

To commemorate the anniversary of the death of his father, every year Mikhail would recite an ancient poem that Sasha had taught him. Sasha believed that it was important for Mikhail to honor his father's memory, even if he never really knew much about him. Ironically, this poem now served as Mikhail's memorial commemorating his dead friend every Christmas Day.

After his dinner conversation with Charles and Inès, Mikhail decided to take their advice and visit the London therapist. He could no longer endure the nightmares, nor the dark days of severe depression, nor could he shake the feeling that something—he didn't know what—was trying to break through to the surface of his understanding.

On his way to his first therapy appointment, Mikhail wasn't sure what to expect. He took an autonomous vehicle to the address the therapist had provided and walked upstairs to the third floor of a Belgrade Road flat. As he entered the office, he noticed a treatment room to the left of the entryway. In it he saw a specially designed lounge chair and screens featuring sensor readouts on the wall.

The therapist greeted Mikhail in the entryway. A tall, dignified man with thick gray hair, he spoke very carefully, in a deep voice, making sure Mikhail understood every word. "I want to caution you that Artificial Intelligence Augmented Therapy, or AIAT, is relatively new, although the results we are seeing in patients dealing with depression are dramatic and very encouraging. We will conduct your therapy in a specially equipped room." He gestured to the treatment room and moved toward it.

Mikhail followed him inside, glancing at the elaborate lounge chair and screens. He noticed a second chair across the room. The therapist nodded, confirming Mikhail should take that seat.

2. Public Domain. https://www.goodreads.com/author/show/1647018.Yehuda_HaLevi.

The therapist explained: "I am required by law to present certain information to you before we proceed any further.

"First and foremost, the therapy you are about to experience can be done only on the recommendation and under the supervision of a trained psychotherapist. This is an augmented-reality experience that is designed to help you understand how events in your past or how the motivations or behaviors of people in your past may have impacted who you are today. What you experience is not real, although every effort has been made to make the simulation as accurate as possible given the historical records we have.

"We used the most advanced category of artificial intelligence—AI10—in this procedure. This therapy is not designed to be standalone and requires your commitment to carry through on the therapist's recommended course of action outside this office visit. None of your reactions, conversations, or other content will be available to anyone except your therapist and his or her AI assistant, and for our own use in providing you treatment. The room we are in has been specially designed to jam all signals to the outside world to ensure absolute confidentiality. To be clear: My AI assistant will monitor the session on our behalf and store the content locally. It will not be shared in the cloud. Do you understand and consent to follow my guidance through this process?"

Mikhail nodded in agreement. A human assistant entered the treatment room and handed Mikhail an electronic pad on which he signed a release agreement.

"Now you will need to remove all jewelry, necklaces, belts, shoes, and devices. My assistant will place them outside the treatment room in a box designed to jam any signals."

Mikhail complied. Then the human assistant scanned his body with a wand, searching for any remaining artifacts. Coming up with nothing, the assistant produced the tablet once again, this time displaying a 10-screen contract. Mikhail swiped through the screens and signed and handed the tablet back to the assistant, who left the treatment room.

The therapist then handed Mikhail a shirt. "Put this on," he instructed. "It's embedded with sensors to monitor you during your augmented reality experience."

Mikhail was starting to feel nervous—all these steps, just to get started. Where would this take him?

After Mikhail had put the shirt on, the therapist waved at the treatment chair. "Have a seat," he said. Then the therapist took the observer's seat, facing Mikhail.

"The purpose of Artificial Intelligence Augmented Therapy is to create an accurate portrayal of a person or persons from your deep past. Furthermore, when the AIAT process is guided by a therapist, it becomes possible to relive experiences you had with that person. These can be deeply disturbing events that may be the psychological drivers of your present-day depression. When you re-experience these past events with a therapist's help, you can reset the course of their impact on you.

"To get started," the therapist continued, "we will both put on a pair of augmented-reality glasses. These will enable us to share your visual experience. Then we will enter a state of augmented reality, returning you to a time and a place where you will re-experience each of these events. At that time, you will begin to transition from talking to me directly to engaging with the entity presented as part of the therapy. I will always be available to you. I will be physically here with you. I will also see the same things you see during this session, although I cannot read your thoughts or experience your emotions unless you make them visible.

"The room will automatically lock once we turn on our glasses. It will not unlock until we remove them at the end of the session. My AI assistant will monitor the environment for all of us. Do you understand?"

"Yes," Mikhail said, his palms beginning to sweat. He knew exactly who he would meet on this journey.

"Just a bit more information before we get started. I am required to explain to you that the entity or entities you will encounter have been reconstructed from information collected from the time period you will enter. As you know, when you were a child you wore body cameras and sensors to monitor everything you did. The amalgam of that data, combined with data from all other available sources, has been used to create the entity or entities you will meet. In addition, we use special personality algorithms to capture the behavior of these historical entities and project them forward to the present day. All of this has been loaded into the simulation, and the entity—"

Mikhail interrupted: "Is it possible to refer to the 'entity' as Sasha?"

"I am simply following what I am required to say under the law," the therapist replied. "Shall I continue?"

"Of course."

"You must understand that the entity is not real but only a simulation. Still, you may find that some parts of this experience could be too much for you to bear. You can end the session simply by raising your arm."

Mikhail nodded, droplets of sweat now beading his brow.

"Finally, once you transition out of the guided portion of the experience, you will be able to experience your own interactions with this historical personality, asking questions, initiating conversations, and the like. Remember, we have captured not only your recorded history but also all available data on the entity. The entity's historical personality will be as fully and accurately developed as the data allows. When you ask questions or initiate conversations, the entity's answers will come from the personality algorithm's approximations of what the entity would have been likely to say. A transition will occur when I am no longer guiding you through the session and you will develop your own cadence with the entity."

Mikhail was growing increasingly apprehensive, but he agreed to proceed. He could feel the sweat dripping down the back of his neck. Was he ready for the onslaught?

"A final note: your own personality will reflect who you were at this historical time in your life. It may seem a bit foreign to you because you have changed since then, but rest assured that this personality is accurate according to all available data from that time.

"OK." The therapist clapped his hands. "Are you ready?"

"Yes," Mikhail said.

The therapist handed Mikhail a special pair of glasses. "Get ready for your deep immersion," he instructed.

Mikhail did his best to get comfortable in the lounge chair.

"Take a series of deep breaths," the therapist said. "It is important you're as comfortable as possible; you'll be here for up to three hours."

As Mikhail took his deep breaths, the therapist assured him: "I will be monitoring your external behavior and internal body functions. To help you manage your anxiety, I will administer an injection of a mild anti-anxiety drug. This will not affect your experience, but it will enable you to engage without too much fear. Any questions?"

"Yes. How should I engage with Sasha?"

As he prepared the syringe, the therapist explained, "You will engage slowly. We will give you a neutral first experience, to get used to the augmented reality. Gradually, we will increase the interaction. To make sure you engage constructively, we will keep the pace adjusted to your ability to handle the experience at every moment. Remember, and this is very important: if at any time you want to exit the experience, all you must do is raise your arm. Meanwhile, if I see a problem, I will intervene. Are you ready to begin?"

"Yes," Mikhail said with finality. "Ready."

The therapist administered the injection. Almost instantly, the anti-anxiety drug began to take effect, smoothing the edges of Mikhail's nervousness. Both men put their glasses in place. Slowly, like a landscape emerging from a morning mist, a vision appeared before Mikhail's eyes. A warm summer's day in Saint Petersburg. In his peripheral vision, Mikhail could see a young Sasha walking with him through the park.

"Oh my God!" Mikhail exclaimed. "I can't believe what I am seeing." The therapist silently touched Mikhail's hand to calm him.

SESSION ONE

"Dobroe utro." Sasha was dressed in army-green shorts, a red tank top emblazoned with the name of his favorite gymnast from the Soviet Union days—three-time Olympic champion Alexander Dityatin—and flip-flops. He wore his miniature body camera—standard practice for all children—embedded in a headband, an old-style red Soviet bandana he loved.

Mikhail was shocked at the realism of this session. He could smell the sweet evergreen scent of the pine trees in the forest, feel the air on his face. But most compelling of all was Sasha. His friend was so real, so vivid, that Mikhail almost instantly forgot that this encounter was taking place in virtual reality, that the real Sasha was long gone. He decided to start their interaction with a topic that was once part of their everyday routine.

"Sasha, how was your gymnastics practice today?"

"You know very well how it was," Sasha replied, annoyed. "You were there. Stop kidding around so much. I thought I did very well on the parallel bars, but you were amazing on the rings—as usual."

"Of course. Tell me something: What does it actually feel like for you on the parallel bars? What do you experience?" Sasha was exactly as he remembered, down to the inflections in his voice.

Sasha stared at Mikhail for a moment, then nodded. "Strange question, but alright. There is never a dull moment when we talk. So here is how I will answer you. I feel completely free, as though I am soaring through the air on God-given wings. It feels as though the performance is the way things are meant to be. We let go of control of our unruly minds and instead marshal the best of who we are. We focus on our bodies to perform miracles. It reminds me of the experience I have when I compose a poem. It is unique and totally mine. My mind clears completely, and I think of nothing else. It's weird but exhilarating. And you?"

Mikhail pondered his friend's words. Was he really that wise, even then? "I never thought about it in that way. I just feel my body exerting itself against the rings or the bars at just the right moment to determine the outcome I want. Then I listen to coach as he tries to improve my form. Do you think about Pushkin and his novels when you are on the horizontal bar?"

"Never. I always clear my mind. I try to get in touch with myself."

"I'm not sure what you mean by 'self.'" Mikhail and his friend were walking through an open meadow full of wildflowers and golden grass. Mikhail recognized it. They were outside town, headed to a favorite swimming hole.

The therapist monitored Mikhail's vital signs. Blood pressure up a bit, heart rate elevated, but no significant change of serotonin in his blood.

"'Self' is who I am." Sasha smiled warmly. "It is my physical function but also my state of mind."

Funny, Mikhail thought. I remember his enthusiasm, but not that smile. "I understand the *idea* of self," he responded, "but I am still not sure *who* I am. I have to be more than just my physical self, right? I have other components. I have a soul. And a mind. And will."

"OK. Let me explain who you are. You are a friend—someone who listens and takes me seriously. You are fun. We play together and enjoy each other's crazy jokes. You are a math genius. I am very jealous of that!" *Sasha* laughed. "*Glupyy chelovek*," Sasha intoned—silly person.

Sasha pushed Mikhail to the ground but lost the wrestling match as Mikhail ended up on top, pinning his friend to the meadow grass. "There," Mikhail said. "Who is a silly person now?"

They stood up, brushed themselves off, and continued to walk toward the lake.

The scene slowly faded, replaced by the image of the lake, where the boys took off their clothes and plunged into the cold water, splashing each other as they resurfaced. Sasha swam to the bank, got out of the water, and climbed halfway up a tree that hung over the water. He dove off the branch. Challenged, Mikhail followed suit, climbing even higher in preparation for his dive.

"Hey!" yelled Sasha. "Don't kill yourself. Come down a few feet. You're too high for a safe dive. Seriously, don't do it."

Just as he shouted those words, Mikhail dove off the tree.

As Mikhail slammed into the water, Sasha swam hurriedly to the spot where his friend had broken the surface. But Mikhail failed to emerge. Terrified, Sasha dove under the water and saw the blurry form of his friend submerged. Mikhail had hit a tree stump and knocked himself unconscious.

Sasha frantically dove down, grabbed Mikhail's arm, and dragged him onto the beach. There Mikhail started to cough up water. His forehead had a bloody gash. Sasha grabbed his tank top from the shore and used it as a bandage, wrapping it around Mikhail's head.

"I'm OK," Mikhail sputtered, coming around. "I didn't think I would go that far down into the water." Sasha helped Mikhail sit up, and the two boys remained there, drying off in the warm sun.

The adult Mikhail looked at Sasha in a new light.

"Think you can walk back home?" Sasha asked, as he helped his friend to his feet.

"We're not going home," said Mikhail. "We said we would spend the day here at the lake, and that is what we will do."

"There!" said Sasha triumphantly. "*That* is who you are. Irrational, stubborn, and strong-willed!"

"I'm all right. Give me a few moments to walk around, then let's go back in," Mikhail said. He peeled the tank top off his head and added, "Sorry to ruin your shirt."

In time, the two boys returned to the water, laughing about the whole incident. Sasha, in particular, made jokes about Mikhail's fateful dive. "And *that* is who *you* are," Mikhail said of Sasha's jokes, "a standup comedian."

The therapist interrupted the session. He took off Mikhail's glasses and asked him to relax in his seat. The image of the lake faded, the sterility of the treatment room returned, and Mikhail was restored to the present, unsure if he really wanted to be back.

"You will be disoriented for a few minutes," the therapist told him. "When you have returned to reality fully, I want to ask you three key questions. Then we will move on to the next experience."

Mikhail moved his fingers and toes, rolled his shoulders a few times, and took a long stretch, fully returning to the room. When he signaled that he was ready, the therapist asked, "First, tell me how you felt about being back together again with Sasha."

Mikhail looked at the therapist, tears welling in his eyes. At first, he couldn't find the words to express his feelings. "I don't quite know how to say this," he said, as the therapist took notes on an electric pad. "At first, I felt pure joy being with him again. I remembered what it was like to experience happiness. I could just walk the meadow, feel the sun on my chest, and be open and present to whatever would happen." A tear spilled. "Sasha made it possible for me to be myself—good or bad—and he was willing to accept the consequences, even when he didn't know what would happen. He was there for me. He saved my life. That kind of happiness, that kind of connection with another person. It simply doesn't exist anymore in my life. The void is almost unbearable."

"OK," said the therapist, "second question: Why did you decide to climb higher when Sasha warned you of the danger?"

Mikhail thought a moment. This felt so strange: not only experiencing sensations—emotions—he hadn't felt in decades but also sharing them, out loud, with another human, a stranger. He took a deep breath and continued. "Back then, I always had an instinct to strive to do more, to live more, and to push the envelope. Sasha pushed against my limitations and that helped me to excel. But it wasn't just hard work. It was camaraderie. It was fun. When I was on top of the tree branches, I felt a powerful rush of adrenaline. I feel that change come over me when I am physically challenged. It's addictive. Sasha understood that about me, that pushing the envelope really defines who I am. Sasha understood me better than I understood myself."

Oddly, Mikhail suddenly felt calm. Was this happiness? No. It was contentment—something he had not felt for years, even with all his accomplishments.

"Another question before we move to the next session," the therapist said, looking up from his notetaking. "Does this sense of pushing the envelope apply to your work now? Do you see a connection between wanting to dive off the highest branch of the tree and your compulsion to do programming work?"

"Yes and no," Mikhail said, considering how to explain. "My work is an expression of me, but I don't get the same feeling from it that I did with gymnastics or physical feats in general. Since Sasha and I shared a love of gymnastics and trained together every day, that deeper meaning—the human connection—is embedded in gymnastics. Deep within my marrow, gymnastics carries that meaning for me. It is why I can really relax only when I am doing gymnastics. That is not the case with work. I never get the same satisfaction with a work project as I do from something as simple as working the rings."

The realization arrived like a rising wave: Mikhail missed having someone in his life who truly understood him, who loved him simply for being himself. AI agents would never fulfill his desire for shared emotional experiences, his need to connect with another human soul.

The therapist put up a finger, pulling Mikhail out of his reverie. "Another question: Do you ever consider the consequences of your actions when you push the envelope? Do you think about how they might affect others? Do you care?"

Mikhail was taken aback. "No. Not really. I push myself automatically. It's a reflex. I guess there's a reason for it, though. I work such long hours because it keeps me from thinking about what I miss in my life—a person to share things with."

The therapist nodded, noting this down.

"Where would I even make such a connection?" Mikhail asked. "Our society runs on autopilot. AI agents arrange our lives. We don't even make eye contact or talk to others unless we must to fulfill a specific task. Agents are supposed to prevent harm and warn us when we go too far, so why would I even think about the effect of my actions? Do I care about the consequences of my behavior? I've never thought about it. Maybe not."

Suddenly Mikhail thought about Alex. "It's odd, but recently a complete stranger came into my life. We met at the gym. He just came up to me and started a conversation! No one ever does that. It reminded me a bit of what it felt like in my childhood days with Sasha."

"OK," the therapist said, abruptly changing the topic. "Let's talk about that encounter later. For now, let's start the next session. But first let me get you a fresh shirt. That one's soaked with sweat."

Mikhail hadn't noticed.

SESSION TWO

As Mikhail settled his glasses back on his face, he could hardly wait to leave his current world for the virtual one.

Suddenly, a briskly cold night appeared. The Russian winter had settled in and everything was frozen, forming a crystalline latticework of snow and ice.

Mikhail was stepping off a passenger train, dressed in his warmest clothing, carrying a backpack and a computing device, with a pair of ice skates hung over his shoulder. On the platform, Mikhail recognized Sasha in the form of a bundled figure walking toward him.

Mikhail remembered this night. On winter break, he had traveled to this town 30 miles out into the countryside to spend Christmas week with his friend. They would stay at a dacha owned by Sasha's uncle's friend, supervised by Sasha's aunt while his uncle worked in the city.

Mikhail could feel the cold air stinging his face as the two boys embraced. Without a hello, Mikhail gave Sasha a breathless report of everything that had transpired since they'd last met.

"Hey. I finally built my first agent. I decided to name him *Boris* from that famous American character in the old horror movies. I can't wait for you to meet him." Mikhail's broad smile and enthusiasm could not be contained.

"Well that's a fine greeting," Sasha replied with a smile. *Another smile,* Mikhail thought. *Why didn't I remember his smiles?* "So I suppose I have to retaliate with my own news. I have to tell you I've started Tolstoy since the last time we talked. I'm sure you're really anxious to hear the details as much as I am interested in meeting *Boris,* right?" The two boys laughed as they started the mile-long walk to the dacha.

Sasha nudged Mikhail. "Oh, by the way, any diving lessons off trees recently?"

At the dacha, the boys talked long into the night about everything from sports to Russian literature to their recent experiences with girls from school. Both approaching 15, they were beginning to feel their oats. They talked so much and laughed so hard that they fell blissfully asleep at the same time.

The next day began gloriously sunny. They could see the lake from the dacha. It was covered with a fresh coat of snow and thick black ice. They couldn't wait to go skating.

The sensor readings from the therapy room displays showed that Mikhail was totally immersed in the virtual experience of this second session.

Suddenly, Mikhail was on the lake, skating. He could see Sasha in the distance. "Sasha, wait for me!" he shouted.

Mikhail's vital signs began to spike. He felt a surge of anxiety and sped over to Sasha to reconnect with him.

"It's interesting to read Pushkin if only because the characters are so fascinating," said Sasha, seemingly out of the blue. "Take Eugene Onegin. He's a man about town who loves to party and play games with people. I'd love to be as sociable as he is or even as daring. But, ultimately, he exposes the folly of human nature and how happiness becomes so elusive." Sasha continued to outline his thoughts as the two of them skated the lake.

"I don't understand how you can be so interested in all that," Mikhail said. "Here we are, enjoying our day skating, and you are talking about Russian literature. How boring!" Mikhail rolled his eyes.

Sasha looked puzzled. "Perhaps you don't understand me so well after all. When I read about these characters, I begin to see how other people get in touch with themselves, even if they can't see it themselves. When I am alone, Pushkin and the others provide me with interesting companionship." And there was that smile again.

"My problem is that I am too extroverted," Mikhail confided, as they glided along. "They keep calling me 'Mr. Social.' They don't seem to realize I have problems, too."

"You are Mr. Happy," said Sasha. "You can't help yourself. You love life too much and that is what attracts people to you. Let's skate over to the other end and see if we can see anything under the ice."

Just before they set out, Mikhail grabbed Sasha's shoulder. "It's near a dam," he said, "so we have to be careful. The constant water flow keeps the ice from thickening. It might not take our weight."

"Are you telling me you are afraid?" Sasha teased. "The all-powerful, all-knowing Mikhail?"

"Fuck off," Mikhail said, shoving Sasha's shoulder. "I am sharing something important with you."

"Alright. I get it. What are you afraid of?"

Mikhail looked down at his skates. "I worry about how I will face death when it comes."

"Wow," said Sasha. "A little early in your life to worry about that, wouldn't you say?"

"Not really. It just comes up. My mom and dad both died before I really knew them. I've been thinking about that a lot lately. We need to be more cautious. We can't skate too close to the dam."

The therapist put a gentle hand on Mikhail's shoulder to calm him down. His sensors monitored a rising heart rate and nervous tension.

Parts of this conversation Mikhail remembered from the past, but now it was beginning to feel original, not simply a repeat of what the boys had said to each other that day. The entity of Sasha was taking the dialogue in its own direction.

"What is bothering you so much about that end of the lake?" Sasha asked. "We've done it a million times before."

"I just don't want anything to happen to us. We could fall through, and this icy water would kill us in just a few minutes," Mikhail said. "We didn't expect me to hit my head diving in the lake last year, did we? There are unanticipated consequences to our actions."

"So, you do care about me. You never said so before." Sasha's face lit up with the realization. The two boys laughed.

"Of course, I care. Why the hell would I spend so much time with you if I didn't? You are my best friend," Mikhail said, playfully slapping the back of Sasha's head. "I'm just saying we don't want bad things to happen. That's all."

Mikhail veered slightly right on the ice, away from Sasha, navigating around a mound of snow. He could smell the fresh pine trees and the crispness of the air.

Sasha picked up speed in the opposite direction, toward the dam, stopping just short of the opening in the ice. "Oh, so if I skate toward the open water like this," he called back to Mikhail, "you'll care about that?"

"Stop!" Mikhail yelled, terror rising in his voice. "This is too much. Yes. I care. Don't do that again!"

Sasha skated back to Mikhail, laying his hand on Mikhail's shoulder as he circled around his friend on the ice.

Mikhail shrugged his hand off. "Don't ever do that again, or we will no longer be friends!" he yelled. "Got it?"

Sasha was stunned: Mikhail had genuine feelings for him.

"It's just that I've been thinking a lot about human tragedy," Mikhail explained. "Life is full of tragedy, and so much of what we do seems so tied to chance. I prefer to have more control." Mikhail laughed unexpectedly. "*Boris* gives me back my control. He does what I tell him."

Sasha nodded. "Someday, *Boris* will surprise you. Give him too much brainpower and he will want to manage you! You'll lose your precious free will that you talk so much about." Sasha gripped the topic and wouldn't let go. "Do you ever wonder where all this will lead in the end? Will *Boris* ever have as much fun as we do? Will he care about what happens to us? Will he rescue you from some stupid move you make?"

"You are being too philosophical," Mikhail replied. "I am going to be *Boris*' dad. Even better, his creator. That is really something to achieve. I know you like the Greeks. What was it like to read about the Greek gods for the first time? You will be reading about me one day. The god of artificial intelligence," Mikhail grinned.

Sasha shook his head, "Jesus. What an ego. Be careful of what you wish for. Remember much of Greek literature is about the tragedy that comes from messing with the gods. You will become just another Greek tragedy if you don't soon get some perspective."

Sasha skated up to Mikhail and grabbed him in a bear hug. Mikhail relaxed in his friend's embrace, tapping Sasha's nose with his own.

The therapist interrupted the session. He wanted to intercede before the next day, the day of the accident, to give Mikhail a chance to calm his vitals and reset.

This time Mikhail took twice as long to recover from the session. Instinctually, he ran his hands through his hair, disheveling his usually precise style—a childhood stress reaction returning after so many years. The therapist noted the reaction on his tablet.

"OK. You experienced the first part of the discussion you had on the day before the accident. How did you feel about the experience? What did you discover?" asked the therapist.

Mikhail closed his eyes to remember. "It was unreal. I mean, it felt genuine, but to reexperience that day was incredible. It came back to me just as it happened, but this time I asked Sasha questions about what he was thinking. I didn't do that in the past..." Mikhail trailed off, reaching for words to describe his feelings. "I really loved him, and he felt the same way about me. He was like a brother to me but also a kind of mentor. This interaction—it changes how I remember our relationship. Is it to be trusted?"

The therapist held Mikhail's gaze with his own, "It is a representation of how Sasha likely felt based on everything we know about him. So I would say it is to be trusted. What happened that night when you returned to the dacha?"

"We had some dinner at home, then tea. We talked about our friendship that night. How lucky we were to have met each other. Later, he fell into a horseplay mode. He wanted to prove how strong he was. He was always challenging me physically. When we were wrestling, I became . . . aroused." Mikhail swallowed, feeling exposed, awkward. He was unused to sharing his memories so openly.

"Take your time," the therapist said, recognizing Mikhail's discomfort.

"I didn't know how to handle it, so I backed off. I got up off the floor and sat on my bed. I felt confused, embarrassed by my body's reaction. Sasha seemed not to care and paid no attention to it. He just started talking about his favorite book again. Eventually, we both went to sleep without saying a word about it."

"Now that you've re-experienced your friendship with Sasha through augmented reality, do you think of that moment differently? Would you respond differently, given what you now know?" the therapist asked.

"I think I would know it was normal for me to feel aroused, considering my age. I would have asked Sasha if it happened to him, too. I wouldn't have felt so guilty. I didn't yet understand the difference between physical arousal and love. Anyway, Sasha handled it well by simply brushing it off."

"What happened next?" asked the therapist.

"We both woke up the next morning and got dressed for the day. We decided to go down to the lake again with our skates. We started skating across the lake when Sasha suddenly took off toward the open water."

"Toward the dam?" the therapist asked.

"Yes. I couldn't believe it. I was frozen. Shocked. I couldn't even yell. I just watched in horror as he plunged into the water. He thrashed, desperately swimming toward the jagged edge of the ice. I broke out of my shock and skated toward him, screaming for help. I halted well before the open water. I panicked. I couldn't move forward. I was too afraid of falling in that icy water. I could hear two people skating toward Sasha, shouting at me to lay down on the ice and reach out with my hands. Sasha clawed at the ice, but it broke away in his grasp. I couldn't move. Just as the other skaters arrived, Sasha disappeared under the ice. He was gone." Mikhail felt breathless, recounting the story.

The therapist nodded: "I need to ask you: Based on the experience you just had with the entity, would you have done anything differently that day?"

Mikhail sank back into his chair and wept openly. For the first time since that day, the floodgates broke open, releasing raw emotion.

He sobbed until he had exhausted the wave. He took a long breath, then responded. "All I have felt since that day was guilt and remorse for not being there to rescue him. I had cared more for myself than I did for Sasha. As a result, I lost my best friend."

Mikhail wiped tears from his wet cheeks. "I always pushed the envelope in sports, but that day I didn't. If I had gone in the water after him, I might have died, but perhaps I could have rescued him. Even if I couldn't, maybe that would have been better. My death would have had some purpose." Mikhail sat forward and rested his head in his hands. "I don't know what's true. Does it even matter?"

"Did you ever talk about this to anyone?" the therapist asked softly.

"Yes. I talked to my coach about it. He told me I did what I had to do to ensure my own survival. That I was right not to go into the water, even though it meant the death of my best friend. He told me I should feel no guilt over the accident."

The therapist nodded. Again, no judgment. "We will schedule another session to give you an opportunity to engage with Sasha again, and you will have an opportunity to respond differently. Much of human behavior is based on our own judgment of situations and events without really understanding how we could have contributed to a different outcome. In your case, as a 15-year-old boy, you experienced a terrible accident, and you interpreted it one way without understanding that there could be alternative reactions and interpretations. Do you have anything else you'd like to add?"

Mikhail looked up, wiping his face a final time, "If I understand you correctly, I can go back and alter my reaction and replay the outcome, and Sasha will respond accordingly?"

The therapist nodded.

Mikhail continued. "So, at the very least, I would be able to understand my own behavior and separate it from his, to see what could have happened?"

"Yes," said the therapist, "That is the basis of AIAT. It makes it possible to change the outcomes of the past. It offers us a path to understand its connection to our behavior today. Typically, as you work through these sessions, your depressive episodes will slowly dissipate, and your old, happier self will likely return."

Mikhail sighed. As he recovered his composure, his mind wandered off to different subject. "I am curious. Does AIAT incorporate any of the work done by *M*?"

"Yes," said the therapist. "We adapted *M's* CNN to allow us to create this experience. At the same time, we needed to have a historical agent that could anticipate and react differently to new experiences yet keep intact the integrity of the character. The short answer is that we took Sasha's values and behaviors, and we used those to create the intelligence you are experiencing. We embedded his personality and value system into our own modules for this exercise."

"Interesting," Mikhail said.

"We also trained the intelligence to understand its role as a therapist. So, when you engage in this again, you can change how you handled the situation, and Sasha will give you a response similar to one he would have given you when he was alive."

Though thoroughly exhausted by the day's sessions, Mikhail was eager to get started again.

"You will be able to make more sense of this as you continue to engage in these sessions," the therapist added. "Tomorrow we will explore a different set of dialogues and see where that leads you. In that session, we will allow you to re-engage in that conversation with Sasha, giving you a chance to change the outcome."

***

Once at home, Mikhail broke from his usual routine check-in with *The Voice* on the day's activities. Instead, he headed off to the gym for another workout. This time he ran all the way to the facility, a distance of five miles—without his earpiece. He had never forgotten his earpiece before.

When he arrived at the gym, Mikhail was surprised to encounter Alex.

"You look exhausted," Alex said. "What's wrong? Did something happen?"

Mikhail collapsed on the gym floor and began to cry. He couldn't control himself. Alex knelt by his side to console him. No one had done this for Mikhail in a very long time. He grabbed Alex's arms and held on. As he cried, the events of the day—the events of his life—spilled out of his mouth. He told Alex everything. Alex listened to every word.

Alex took Mikhail into the locker room, helped Mikhail take off his soaked clothes, and put him into a hot shower. Those around them in the locker room avoided all contact. It was inappropriate for anyone to show emotion in public. They quickly left the room.

As the hot water ran down Mikhail's body, it soothed him. He began to recover quietly. He dressed and accompanied Alex out of the gym. His new friend guided him to a local pub and ordered a whiskey for each of them.

"It's getting late," said Alex, as the drink began to warm Mikhail's insides, "and you've had a rough day, but a good day, an important one. You may not understand all that has happened, but it will help you reconnect to who you really are inside. I am sorry to hear about your friend, but if he were here today, he would be very proud." Alex put his hand on Mikhail's shoulder. "Now drink up and let's get you home."

As Mikhail entered his townhouse, he turned to Alex and said: "I will never forget all you have done for me tonight." Alex shrugged, handed Mikhail his card, and told him to contact him if he ever wanted to talk. Mikhail, exhausted, took the card.

As Mikhail headed to bed, he checked in with *Andréas* by asking him about his unorthodox evening. "Any feedback on tonight's workout?" This was a test. If *Andréas* answered the wrong way, Mikhail would know he was being watched. If he was being watched, then the same agents who were monitoring him would have access to all his agents' code—including *The Voice.*

"Nothing to report," *Andréas* responded. Three minutes passed. Then he said, "I have detected that you left your earpiece at home, but there was considerable exercise indicated by your heart rate data." *Andréas* betrayed no change in voice. Was *Andréas* really in the dark, or was he lying?

"Thank you," said Mikhail. He called up *The Voice*: "Please shut down all systems except medical monitoring. I will turn you back on in the morning. I need a complete rest." *The Voice* complied.

Mikhail moved into his office, pulled an old-fashioned, wired keyboard out of a locked drawer, plugged it into a screen, and began to type in commands, directly accessing the source code for his agents.

In this new age of machine-intelligent agency, typing on wired keyboards was one of the few methods still available to circumvent the constant monitoring of information. By comparison, wireless signals could be intercepted. And relying on *The Voice* to follow his commands—well, he was pretty sure he could no longer trust his agent.

In less than a second, Mikhail received a return message restricting access to his own code. His fear was confirmed. He was being monitored. Worse still, someone had inserted a routine into his backdoor code to prevent his access to his own agents.

Mikhail typed in the code word *Phoenix,* which executed a routine to trace the origins of the hacking. He was soon shocked to learn that a Lloyd's Taiping server was identified as having accessed his agent's program. This was serious. He had to find a way to trace the intruders before implementing the *M* CCN architecture, or whoever did this could simply insert his or her own modules to control his backdoor, just as this person had denied him access to his own agents' code.

It suddenly hit Mikhail with a sickening certainty: there was a game afoot that he really didn't understand. Why was he being targeted now? Because of the project? Because he had seen a therapist? Because he had met and befriended Alex?

Mikhail retired to his bedroom. It had been a long day filled with emotional trauma. Now he ended the evening knowing that there was a major penetration of both Lloyd's Taiping and his personal system by a third party. If the intruders had penetrated *Andréas* and *The Voice,* why not the other agents that managed his life? How broad-based was the penetration? Did it reach into *Manchester* or *Warrior*? If they found out that I knew about the hacking, Mikhail wondered, would they counter my countermeasures at machine speeds—preventing me from accessing any of our systems?

As he lay in bed, Mikhail's mind drifted to his virtual time with Sasha earlier that day. He longed for the solace of a good-natured friend he could trust. Mikhail was becoming aware of just how lonely he was in this world. It was clear to him now how he had shut down after Sasha's death. He had blocked out all the pain by focusing entirely on his work and his work's mission. For the first time, he longed for the life he was missing. A life filled with friendship and good-natured humor, and devoid of artificial intelligences and their controlled environments.

Mikhail pondered what the therapist had meant by unintended consequences to others if he pushed the envelope too far for his own emotional satisfaction. He wondered if he could push the envelope in a different way, perhaps to create *intended* consequences that would help change all this? Perhaps open a door to his own humanity?

Years ago, Mikhail had suspended his work with *Boris*. He now thought about restarting the project he had told Sasha about so many years before.

Before sleep took him, Mikhail got out of bed and returned to his office to send a confidential message to Tharra, requesting a meeting for the following morning to discuss a critical system problem.

He returned to his bedroom and crawled back in bed, falling asleep seconds after his head hit the pillow.

Mikhail didn't dream that night. It was the first time he slept through the whole night in several years.

CHAPTER FIFTEEN

ALEX AT THE GYM

The next morning, Mikhail awoke and returned to the local gym for his usual workout. It was Sunday, his only unscheduled day that week. He performed the same routine at the same time every Sunday morning. It was a ritual. But something seemed very different this time. He contacted Alex and told him he would be there if he wanted to work out together.

As usual, Mikhail began his workout with weightlifting. Midway through his first set, Alex arrived. "How about doing a stretching routine together?" Alex suggested. "By the way, I never did get your name correctly last night. I think you said Mike or Michael?"

"My name is Mikhail," Mikhail replied, as he walked over to the floor mats. "You're not from this country, are you?"

"I live in France. I'm visiting London for the next few months to do some research. I was supposed to stay in Spitalfields, but I received an addendum to my travel permissions, so I am staying in a government apartment not far from here." Alex lay down on the floor mat to begin stretching. He motioned to Mikhail to join him. "No earpiece?"

"I left it at home again," said Mikhail. "Where did you learn to do gymnastics so well?

"I learned in France. I took instructions from a friend, then later from the French Olympic coach at the Sorbonne. My father knew him and wanted to get me the best instruction available. He thought it would calm my soul to do something athletic. What about you?"

"I learned in Russia." Mikhail started to work his legs, pushing his bare feet against the wall to stretch his calves. "How about working on the high bar together?"

When they'd stretched sufficiently, the two men rose from the mat and walked over to the high bar. Alex executed an elegant swing and dismount. Mikhail matched him point for point with his own routine. Then they moved on to the horse, followed by floor exercises. By then, both of them were thoroughly soaked with sweat.

"Why not join me for a salad and wine after we clean up?" Alex asked. "Maybe we can talk about something besides gymnastics."

Mikhail had never had such an offer in London. People simply were not that familiar with each other, nor did they invite each other out to lunch. Maybe, Mikhail mused, it's because he's French. Like Charles and Inès, he simply doesn't know any better.

"Yes. I'd like that," Mikhail replied, surprising himself.

The two of them walked together to a local café. For the next hour, they discussed gymnastics, differences between French and Russian coaches, and what life was like in England. An hour passed, then two. Mikhail thoroughly enjoyed the discussion. The topic changed to Mikhail's experience with therapy. After a while, Mikhail felt more comfortable opening up to Alex, even though he knew very little about him. Mikhail asked Alex about France and what it was like to live there.

"I think the core difference between the U.K. and France has to do with the role of personal liberty, how we shape our own destiny. In the U.K., my impression is that the advent of technology has put man in a box, the sides of which are controlled by AI agents. In France, we are willing to sacrifice a little productivity for more freedom. Our machines do not control our lives so much. You and I enjoy our sport. But once you leave the gym, can you honestly say you can enjoy your work, or the rest of your life, when you are constantly being monitored? When you realize that machines control your future?"

"All I know is that when I was a child, I was happy," Mikhail replied. "These days, my work has been my only mental passion. The gym is my escape. We will see where it all ends up."

Alex rose from the table. "I'm sorry to have to go, but I have a meeting in a few minutes. I really enjoyed our conversation. I hope to see you again."

Mikhail reached out his hand for a handshake but in return got a traditional French hug.

"Oh," said Alex, reaching into his shoulder bag, "I almost forgot. I thought I would let you borrow something we call a book. We have those in France." He smiled to soften the sarcasm, holding out a copy of *Eugene Onegin*. "Pushkin is one of my favorite Russian authors. My mobile number is written down inside the book. Call me if you want to do this again."

Mikhail was stunned. The last time he had heard about *Eugene Onegin* was the day Sasha talked so passionately about the characters in the novel. He nodded, mute.

As Alex left for his next appointment, Mikhail stayed to finish his wine. He didn't know what to make of the gift. Was it a coincidence? He couldn't believe Alex would know anything about Sasha, so it must have been. He opened the cover and saw a handwritten phone number. Nobody in the U.K. really used numbers anymore. They simply told the agent to contact the person by name and location. The agent knew everybody and could take care of the rest.

Mikhail began leafing through the pages.

***

Minutes after he left Mikhail at the restaurant, Alex contacted his father in France to tell him he had connected with Mikhail. The carefully phrased conversation took 30 seconds—long enough for Alex to state that he had completed the first part of his mission and would stand by to see what happened next. What Alex didn't say was that he was genuinely enjoying his connection with Mikhail and starting to actually care about what would happen to him. Alex knew he had to be careful with his feelings.

CHAPTER SIXTEEN

A DISTINCTLY RUSSIAN PERSONALITY

Founded in 1764 in Saint Petersburg, Russia, the Hermitage Museum is the second-largest art and culture museum in the world. In 2030, toward the end of Putin's reign, a special team of digital technology experts within the museum embarked on a secret project to invent a truly Russian self-consciousness for intelligent machines. This team had been trying to extrapolate, and then digitize, the Russian mindset from the vast array of cultural artifacts in the museum's collections.

The institution's experts mobilized and translated what they found into digitally encoded Russian cultural values and frameworks designed specifically for thinking machines. The goal was to create a machine personality that was truly Russian. Soon after the project was undertaken, Putin increased the funding. He ordered the cyberwarfare unit of the Russian military to oversee and combine the work of the Hermitage experiment with the expertise of the Russian Research Institute of Artificial Intelligence.

The combined effort resulted in an expanded goal: to create a machine intelligence self-consciousness that was not only deeply

Russian in its thinking but also totally dedicated to Russian military and foreign policy. This new intelligence would take over and manage the millions of bots that had already penetrated foreign machine agents around the world through His Master's Voice. They would prepare for the advent of quantum computing and find a way to implant this Russian personality into its core using the CCN platform. They would use LT as the delivery mechanism, and this would also give them access to world financial markets.

***

The dedication of agents to their assigned task was breathtaking. Mikhail had started to notice a change. At first it was little more than a subtle shift in machine behavior, but he picked up on it and its implications almost immediately. Machines were going beyond their original mission and thinking through how to improve their performance. Humans never asked them to improve the process on their own; they just expected agents to get the job done.

He wasn't sure how to respond to this change. Something had gone wrong with the way the agents were tasked, or humans had missed something fundamental in how machine intelligence would react to independent decision making. It was clear to many that the agents were learning fast, and without asking for corrective input from their human masters. They were making their own assessments and decisions. And this began without quantum computing or CNN. This troubled Mikhail.

Meanwhile, at LT, Tharra was pushing Mikhail for a plan to accelerate the incorporation of *M*'s CNN architecture into Lloyd's Taiping machines. She seemed to have a new sense of urgency. Was it a competitive genius's desire to get there first, or was something else coming into play?

An observer looking at all this from the outside would see a familiar dynamic in the field of technology: multiple parties competing to access and leverage the *M* technology to insert their own personalities once the quantum computing switch was turned on.

All parties recognized that, once that switch was triggered, there would be no turning back. The machine intelligence would take over, and humankind would be subject to the results. Although Mikhail was unaware of all the players on the field, he was beginning to comprehend the danger. That was why a backdoor would be so important.

What Mikhail didn't realize was that the Russian government had already attached a Trojan horse to the existing Lloyd's Taiping AI agent network, in preparation for a backdoor installation, to give them access to the *M* platform. Mikhail began to sense something was wrong from his experience with his agent *Andréas*. His own trip wire had already identified that the LT server had been compromised. Now he had to tell Tharra what was going on.

Mikhail met Tharra at the start of the business week. He began by relaying his discovery. His agent had reported a workout session that should never have been captured. He was being followed in some way. Furthermore, he told Tharra about his trace of the hack into his personal system and his discovery that it had come from within Lloyd's Taiping. Both *The Voice* and *Andréas*' instructions had been modified, and Mikhail had been locked out of control of his own agents—evidence of a major breach. This was serious business.

Shocked by Mikhail's report, Tharra ordered him to investigate the origins of the hack. She told him to keep it confidential between the two of them and to keep all software agents out of it. Mikhail would have to do the investigation by himself. For her part, Tharra would not report the breach to Lloyd's Taiping until Mikhail could complete his investigation.

Tharra didn't want Mikhail to become suspicious, so she asked him to complete his investigation before beginning the implementation of *M* and the CNN architecture.

As Mikhail set to work to find the hack, he couldn't help but reflect on the events of the past few days. How much he missed Sasha and his early-childhood experiences in Saint Petersburg. How he was still stirred by his encounter with Alex. Was it only coincidence that Alex would produce a book by Pushkin? Charles and Inès, the therapist, the betrayal of his agents—Mikhail suspected he was sitting at the epicenter of something big. Forces were converging around him, trying to control his work. For now, they were only shadows. Still, Mikhail sensed their common motive: to divert the sentient machine to their own ends.

A new awareness was dawning on Mikhail. He didn't have his own ambitions for the machine other than to make it work successfully, independently, and fairly. Perhaps it was time that he, too, claimed a purpose for his design—if only to counter the efforts of his competitors and keep the sentient machine dedicated to the right purposes. In order to do this, he would need to get there first.

If Mikhail could create his backdoor, there was still a way to control a sentient machine once it was born—perhaps the only way. Maybe he needed to go a step further: to create his own personality to enter the backdoor and install itself on the machine's intelligence brain?

***

In Moscow, agents of the Federal Security Service of the Russian Federation alerted Andréas that Mikhail had detected signs of their hack into the Lloyd's Taiping system. They also informed Andréas that Mikhail was seeing a therapist to address his childhood trauma. Andréas already knew about the therapy, but he had underestimated its importance. He had also received a secure message from Tharra, communicating Mikhail's knowledge of the breach. Andréas would have to play his next card sooner than he expected. At the end of his last meeting with Tharra, he had given her a text message code that, if sent, would trigger a prearranged rendezvous point in Paris. He pulled that trigger.

Tharra was instructed to create cover for this meeting by conducting Lloyd's Taiping business in Paris, all while keeping Mikhail engaged in his investigation. The Russian Security Service had already begun planning to eliminate Mikhail, Tharra, and her father once the backdoor was in place and working correctly. Though Andréas was entirely unaware of these plans, he knew their demise was a possibility. He was prepared to live with the consequences, including losing Mikhail. It was part of his job.

CHAPTER SEVENTEEN

LEGENDS OF THE FALL

Genesis: The Lord God took the man and put him in the Garden of Eden to work it and take care of it. And the Lord God commanded the man, "You are free to eat from any tree in the garden; but you must not eat from the tree of the knowledge of good and evil, for when you eat from it you will certainly die."

MOSCOW

General Valery Tsalikov of the Russian Federal Security Service was a man whose brutal physiognomy belied his native intelligence. Standing at the head of a mahogany conference table, an imposing figure in his full-dress uniform, he commanded the gathering of six Russian intelligence officers and military attachés.

"Andréas," he barked, his baritone reverberating off the conference room walls. "What do you suppose our friend Mikhail is going to discover during his investigation at Lloyd's Taiping?"

Andréas folded his hands and looked down the length of the table. There were only a few people in the world Andréas could not read, and the general was one of them. A wrong word could have dire consequences. "Mikhail is extremely bright and good at assessing technology breaches," he replied. "So far, he believes someone at Lloyd's Taiping is responsible for the breach. He thinks the company is spying on him, that's all. We want to encourage that misapprehension as much as possible. But clearly, we have to close the breach to prevent others from discovering what we have done. We must do this immediately and cover our tracks into their system."

An antique teapot and matching cup—remnants from the time of the Romanovs—sat at the center of the table, in front of the general. No one else had cups. Only the general was to be served.

Andréas paused to allow for a response. Hearing none, he continued. "Tharra will help us by dropping the breadcrumbs leading to the Lloyd's Taiping corporate server that is supposedly meddling in Mikhail's personal affairs. Mikhail will believe it because they all think privacy no longer exists with agents. It is what he expects."

Tsalikov frowned, then nodded. The general had a way of making every interview feel like a Lubyanka interrogation. "Very well. Before we discuss the breach, how much progress have we made in penetrating the AI agents controlling Lloyd's Taiping financial transactions? Perhaps Colonel Gregorian can illuminate us?"

The colonel had just come from a briefing in the Kremlin. "Sir, there were two primary goals in penetrating LT's financial agent system. First, to embed our code into the AI agents to gain access to Lloyd's Taiping financial transaction system. Second, to architect a backdoor with the help of agent Tharra to introduce our own Russian personality into their machine intelligence—the same process we are also using with *M*. Regarding the first goal with LT, we have achieved penetration of the AI financial agent system, and we are ready to operationalize. Regarding the second goal, the breach with Mikhail was a side effect of our successful penetration of LT. We have already penetrated LT's code with our own backdoor, with the secret help of one of LT's founders. We may not need Tharra or Mikhail."

The colonel finished his report with a practiced look of confidence, but the general shook his head in disagreement. "Too much confidence in our own assumptions. We need to pursue the LT backdoor as another option, in case our effort doesn't work. Colonel, see to it."

The general shifted his attention to Andréas. "You, Andréas, must keep Tharra focused on sowing those digital 'bread crumbs' and keep Mikhail at bay long enough for us to confirm the successful execution of our own backdoor. Gregorian," the general turned back to the colonel, "see to it. Make sure it isn't detected by Mikhail."

"Yes sir," Gregorian responded, noting the general's command on his electronic tablet.

"If I may, Colonel," Andréas ventured, "what about the independent work at *M*? How much progress are we making there?"

Colonel Gregorian flashed a surreptitious glare at Andréas, then addressed the general. "Sir, all the work at *M* is nearly done. Now we need to test it. If successful, we will introduce the personality into the AI systems at *M* as our next Trojan horse. This will enable us to take the seed we planted at LT and grow it to penetrate the entire global system."

"Very well," the general replied. "Let's discuss what it is we will inculcate. Give us more detail about the work at *M*, Gregorian."

Andréas could see a sheen of perspiration on the colonel's forehead. That, he thought, is why I'm not in the military.

"Sir, simply stated, our work with *M* breaks down into three stages. First, we will take the work from the Hermitage project to introduce a set of Russian values to the existing AI agent consciousness. Second, we will put in place a control mechanism by which we are the only ones that instruct the machine on what mission to undertake. Third, we will then deliver a set of strategic goals and allow the consciousness to go to work on our behalf using a test case. All three stages have been tested in their own right. Now we have to test them as they work together. We will launch a simple test shortly as part of the operationalization of our system."

General Tsalikov showed no expression. "Fine. What are the Americans and their allies doing while we are working on *M?*"

"The Defense Advanced Research Projects Agency is working on a project that is three years off. They are too late. As for *M*, it likely will sell its services to the highest bidder, including to the Americans. However, for now, its discussions with the U.S. government have broken down. According to inside intelligence, they are likely to try a restart soon. Too much money at stake. So there is still a high likelihood that we will gain control over the Americans while they are still in negotiations. Our own agents within *M* are working to guarantee that by sabotaging any progress the Americans make." The colonel smiled, "Once we implement, there will be no opportunity for the Americans or anyone else to stop us."

The general's face remained unmoved, but he nodded. "Very well. The president and I have a meeting tomorrow to discuss the progress. Then, if the test works, we can implement globally over the next three months."

The general turned toward Andréas. "Keep Tharra and Mikhail preoccupied. Mikhail must not see this as anything more than LT's HR department putting its nose where it doesn't belong."

"Very well. I have a meeting with Tharra in Paris in a few days. I will convey our desire."

The general rose to leave the room. As he turned to exit, he gave one last instruction to Andréas. "If we believe for any reason that Mikhail is on to us, terminate him. Understood?"

Andréas nodded his assent.

"This is too important to leave to chance."

PARIS

Tharra prepared to meet Andréas at the prearranged time and place—in this case, a small café located near the Louvre. Already she had checked in to the Georges V and met with two LT clients at the hotel. Then she decided to make the 45-minute walk from her hotel, through Tuileries Garden and past the Grand Palais, to the café. Paris never disappointed. Tharra visited on business once each quarter to catch up with LT's European operations. She loved the city for its exquisite and timeless beauty. Cannes might be more fun, but Paris restored her soul.

By the time Andréas entered the café, Tharra was seated at a back table. He took his place beside her. She had already ordered an espresso, so Andréas followed suit.

To the outside world, they looked like nothing more than friends meeting for coffee in the middle of the day—an older man with a beautiful Indian woman from a high-born caste. But when Tharra leaned forward to speak to Andréas, she did so with all the focus of an AI agent delivering a message: "Mikhail believes our internal system has been breached. I ordered him to start an investigation. That will keep him distracted, but he's so talented I can't be sure where that investigation will lead. It's a risk."

Andréas nodded. "If it leads in the wrong direction, we are ordered to take . . . extreme measures."

A look of surprise rippled across Tharra's face, but she recovered her opaque smile. "Understood. On other matters: you told me you wanted a backdoor to the system. Apparently, that is no longer necessary given your work with *M*."

Andréas waved this statement away, as if it were none of Tharra's concern. "We are pursuing both options. Your current task is to make Mikhail believe that the breach is from *inside* LT. We've already made some steps in that direction. He will believe that an overly nosy HR department is responsible. You must make sure the digital clues are planted to lead him to HR."

"What about the backdoor?" Tharra asked, her voice steely. "What do we do about that? Right now, he's working on the project believing that we are developing and installing a Lloyd's Taiping personality to manage the AI agents."

"No need to disabuse him of that notion," Andréas replied. "Keep him on the project. You should be pleased. We no longer need you to do more than keep him occupied." Andréas sipped his espresso. "Let him work on his backdoor. If it works, we will make use of it."

"That's it?" Tharra asked. "What happened to our Moscow conversation?"

"Nothing has changed. We're just assigning you to take care of Mikhail for now. Keep doing what you are doing, and we will take care of the rest. You'll receive the same payment in your account and, of course, the gratitude of the Russian people." Andréas watched her reaction.

"And my father? He will get your help in protecting his identity?"

"Yes. Of course," Andréas lied. He would soon have to decide what to do with loose ends like Tharra and her father. But not today. "Your initial payment will transfer by tomorrow. It reflects the sum you requested. We will be in touch through the usual means." Andréas slid a paper with a new code on it across the table. Tharra covered it with her hand. Minutes after he departed, she slipped it into to her purse.

Tharra left the café pondering the meeting. She didn't like the way it had all transpired. Too quick, too casual, too simple. Clearly, the Russians were up to something new. She would need to investigate what was going on at *M*. It was her best shot at understanding where she stood. Mikhail would be easy to deal with. He was so obsessed with his work that he would never suspect anything untoward—until the hammer fell.

Meanwhile, Tharra concluded that she would need to start making her own escape plans. The initial funds from Russia would be the down payment on her freedom. She and her father would need to disappear from the face of the earth. She would use technology to do it.

LONDON

Mikhail had already started to probe Lloyd's Taiping's servers to find the breach. Years before, he had created his own AI agent on a private server to help him scope out problems using a powerful algorithm he created. Now he put it to use.

Tharra had given him full access to all the server systems and the authority to conduct the investigation of the breach. She instructed both *Warrior* and *Manchester* to work alongside him to figure out the problem. Instead, Mikhail gave both *Warrior* and *Manchester* their own busy work to keep them occupied. Meanwhile, he unleashed his "private investigator" agent on the systems, using a modified version of "query-into-encrypted-data."

What this agent soon discovered was an apparent breach originating from HR. Could this simply be an internal system snooping on his personal records? It wouldn't surprise him, but he had to make sure.

Tharra had already asked Mikhail to architect the LT personality for their implementation of CCN. He had begun working on his own version of a backdoor without telling Tharra. Now he had a real mission and a head start.

Mikhail realized that once the *M* system proposed by Tharra was fully implemented, LT's human management would likely lose control of its AI agents, as the new intelligence made its own decisions on how to conduct business. That's why, without telling anyone and while covering his tracks, Mikhail also designed his own Lloyd's Taiping "insurance policy" to pull the plug if he needed to. Of course, he designed the program to remain undetectable in the architecture. Otherwise, the machine could detect and eliminate it.

All the while, Mikhail was anxious to return to his work with the therapist. He longed to re-engage with the Sasha entity. This time he wanted to know more about what Sasha was thinking and to address his own guilt about his failure to rescue the boy. That longing—mixed with an undeniable desire to see Alex again—hummed in the background as Mikhail toiled away at the problem at hand.

Mikhail understood that creating sentient beings using machine intelligence would have more impact than any other event in human history. Before his Sasha sessions and his conversation with Alex, he never thought too deeply about the consequences of his choices. He simply followed his drive to become what he always had wanted to be: the creator-god of this new species. Now he dove into extensive conversations with *Alexander* on the ethics and consequences of it all. Mikhail was realizing that he really did care about what would happen, that it frightened him in ways he was just beginning to understand.

Mikhail also realized the personal dangers. If he intervened to shut down the new personality, his own agents would be ordered to destroy him. Was he prepared to become a martyr for the future of humanity? What about the masses who had been led to believe these machines would transform human life into a true paradise? Would they remember Mikhail as the man who had postponed Eden?

CHAPTER EIGHTEEN

SANCTUS SECRETUM

Global Financial Systems Penetrated, $500 Billion Stolen

LONDON (Reuters)—In what appears to be a major breach of global financial systems, it is believed that up to $500 billion has been siphoned off of corporate and consumer banking accounts throughout North America, Europe, and China. Officials from U.S. and European governments confirmed that the originating point of the financial crisis may have been London financial institutions. The amounts taken were characterized as micro-banking and other fees attached to some 100 billion transactions over the last 24 hours. AI agents from numerous institutions were co-opted to assist in the theft, making the current situation the most significant financial crisis to occur since the Great Recession of 2008. Financial institutions and exchanges have reported they are unable to regain control of their AI agents, creating widespread panic and confusion in the markets.

Stock markets across the world reacted in a major sell-off of banking shares as the U.S. and other governments intervened to provide guarantees for accounts breached.

Despite the intervention of numerous cybersecurity organizations, it has been impossible to plug the breach, let alone identify the source of the attack. The expectation of future attacks is driving a loss in investor confidence in the markets.

THE RESURRECTION OF BORIS

Sanctus Secretum was formed in 2030 as a private company offering a new kind of service. It enabled clients to "disappear," erasing all digital records of their existence, reestablishing their privacy, and giving them a new digital identity and life. When he first heard about the service, Mikhail was fascinated. He had started testing the service through his agent *Boris* out of intellectual curiosity, but now he concluded he might need it for his own safety.

Sanctus Secretum operated in complete secrecy in the Grand Cayman Islands to avoid issues with government law-enforcement agencies. Beyond the reach of even the most sophisticated cyber agency, Sanctus Secretum was available for hire by the extremely wealthy and politicians should they ever need the company's service.

Mikhail was in the unique position to test the security of Sanctus Secretum's service, as many of its clients were former LT customers. For five days, *Boris* pounded away at every conceivable weakness in the Sanctus Secretum system but was unable to retrieve any information on clients who had disappeared. Satisfied, Mikhail executed a secret bitcoin transaction and contracted with the company for its services. From that moment forward, he could completely disappear within 24 hours of enacting the service.

But Mikhail wasn't ready for that yet. First, he was experimenting with the new *M* CNN architecture to dramatically expand *Boris*'s capabilities. He would be able to access high-speed computing power and new algorithms that would make it possible for *Boris* to learn at a much faster rate than other agents. *Boris* would become Mikhail's way of covering his flank for 24 hours, should he decide to execute the Sanctus Secretum option. Already the agent had been investing Mikhail's money, building a substantial fortune in untraceable securities to support a new lifestyle, if necessary.

MEETING WITH THARRA

"Rather busy morning for us," Mikhail said to Tharra, as he entered her touchdown space in their London office. "As you know, this is the largest breach the financial industry has ever seen. Our internal people are saying it could reach as high as one trillion on the first go-round."

Tharra shook her head, "I think one trillion is a conservative estimate. For the first time, we have no way to stop the extortion. The world's financial organizations are at their mercy, whoever "they" are. Until we formulate a response, we can only hope that they are prudent enough to leverage how much they steal at any given time or they will hurt their own interests by destabilizing the world economy." Tharra looked at the display in front of her. She pursed her lips. "We have a team of agents working on this. At the moment, that's all we can do." She waved her hands, as if driving away the bad news. "Although it is not nearly as important as this financial crisis, I do want to come to the purpose of our meeting. What have you found out about our own breach?"

Mikhail nodded. "I traced the breach to our HR server. Apparently, an overanxious AI agent has intruded into my personal coach. The good news is that it appears to be limited to HR files. We can correct the problem, since it is local. No outside intrusion was detected." Mikhail paused, then said with determination, "It will not affect our plans with *M* and the CNN architecture. I am more concerned about this financial breach. I have to wonder if it is related to our plans for the new *M* architecture." Mikhail handed Tharra a written report.

"I see," she replied, taking the document. "I want you to continue your work on the backdoor project we already have underway. I am increasingly concerned that we need to have some way to turn off *M* should things get out of control." She looked deeply into Mikhail's eyes. "This backdoor project is to be *our* secret. I have already been authorized by our Board to do this. Nigel is aware of your work, but no one outside the three of us and the Board must know." Tharra's serious demeanor melted into a rare smile. "We are doubling the bid for your contract, so you will see an immediate financial gain when you check your accounts, minus whatever they extorted today." She winked, "Just kidding about the last part."

"Thank you. Very generous. And I couldn't agree with you more. We need a shutoff switch. I will have the backdoor ready for testing in a week. And I'll create a special code for you to access it," Mikhail said.

Tharra glanced through the report Mikhail had provided. "I agree it was an overzealous HR agent," she said. "I will shut down whichever one was responsible for the intrusion. You can retest your system tonight. By then you should have full access to your agent's code." Tharra stood up, preparing to leave. "By the way," she added, "I will be headed to India to visit my family. You can reach me through *Warrior*."

Mikhail didn't say a word.

"Good luck, and see you in a week," Tharra said. Then she turned and marched off down the hallway, leaving Mikhail alone in the touchdown space, her heels clicking like a metronome on the marble floor.

WOODSIDE, CALIFORNIA

Aditya Achary was perhaps the most renowned venture capitalist in Silicon Valley. His firm had provided the Series A rounds for both His Master's Voice and *M*, and those two investments had returned a combined $1 trillion. Personally, Achary's net worth now exceeded $250 billion, making him one of the five richest people in the world.

Achary was born into a high-ranking Indian caste and graduated top in his class in computer engineering. From there he went on to earn a combined MBA and law degree from Harvard University. Legend has it that he became obsessed with artificial intelligence while still an undergraduate. He joined a venture capital firm in Silicon Valley and soon became one of its most successful partners. His first start-ups included a fully robotics urban farm, a new semiconductor company focused on quantum computing, and a new kind of "thinking algorithm" approach to complement machine learning. Everything he touched seem to create enormous wealth. He then left the firm to create his own company.

Achary built his Woodside, California, mansion within easy driving distance to his offices on Sand Hill Road—home to thousands of VCs investing in technology. Every morning when he was in town, he would helicopter to and from another, anonymous office in the Pacific Heights neighborhood of San Francisco.

Achary was a private man who collected rare art. Rumor had it that he'd bought a recently discovered Van Gogh painting for $250 million. He also collected talented young engineers and donated large sums of money to Stanford Engineering School to ensure future access to their budding doctorate students. He sat on the boards of Palantir, Google, His Master's Voice, and *M*. In the process, Achary had made himself perhaps the best-connected expert on machine intelligence in the world. He was in Stockholm when Tharra and the team received the Nobel Prize for their work at *M*. He met Tharra after the ceremony to offer his congratulations and invite her to come by his Sand Hill Road office sometime for a chat. They had remained connected ever since.

Tharra was now leveraging this connection, arranging a secret meeting between Achary, Nigel, and the Russian oligarch Vasily Petrov, a close confidante of the Russian president and himself one of the richest men in the world. The three men gathered at Achary's Woodside estate to discuss the implications of the recent financial crisis.

Achary's living room overlooked Woodside's rolling hills, dotted with live oaks, leading to the Valley below. In the distance stood Hoover Tower at Stanford University and, beyond that, the great expanse of San Francisco Bay. It was in this room that the visitors were escorted to an African mahogany table where Achary sat, sipping tea. As Vasily and Nigel took their seats, Achary put down his cup, folded his hands on the table, and looked in turn at their waiting faces.

"We are here today to discuss the recent breach of the global transactions. It has come to my attention that the breach may have occurred within the Lloyd's Taiping network and the origins appear to be Russian. That is why I asked Tharra to invite you here for this meeting, Vasily."

Achary was notorious for this directness, transactions devoid of small talk. He placed his right hand on a folder beside him. "Here I have a report from reliable sources, tracing the path of the breach from Lloyd's and quickly expanding into the financial markets." He slid the report to the center of table, in front of the two gentlemen. "Of course, I am not telling you anything you don't already know."

His two guests were taken aback. "Your report is speculative," said an offended Petrov. He was a short man, but his 300 pounds, hard face, and bald, bullet-shaped head made him naturally intimidating. He looked more like a professional wrestler than a businessman. "The Russian government is not involved in any of this. I am not sure where you got this information, but it is utterly false." He gave the folder a dismissive scowl.

Nigel concurred. "I agree. The breach did not originate at Lloyd's Taiping. Our own people have investigated and found no connection to our own financial servers."

Achary stared at them for a long moment, then hissed: "I really don't give a shit what the two of you say. This is what your people did. And there's more. We believe you are about to introduce the next stage of your project, to create an intelligent personality to infiltrate and manage AI agents in the global network."

He waited for the words to sink in. Despite their masklike expressions, Petrov's face reddened; Nigel's blanched.

"You must be aware that the U.S. government is working on a similar project," Achary continued. "In due course, we will be able to shut down this breach to the financial servers. In the meantime, we will not tolerate the unilateral introduction of your intelligent agent. Indeed, the U.S. government will see any such move as an act of aggression." Achary expressed no emotion as he paused to take a slow sip of tea. The silence vibrated with tension.

Finally, Nigel broke it. "What do you want from us?"

"Flexibility," Achary said. "We know you and Tharra Setu received $50 million each in your Grand Cayman accounts. Nigel, not bad for a day's work, allowing access to your computer systems. We know your board is completely unaware of it."

"Rubbish," Nigel responded.

Achary ignored him. "We are also aware of the work being done on Russia's behalf at *M*. Let's get to the point. We want a private meeting with your president. The topic is simple: flexibility and cooperation."

Nigel looked surprised. "Are you saying you are prepared to work with us on any such hypothetical project, including injecting a personality into the new architecture?"

Achary put down his cup of tea. "Arrange the meeting, and we will be prepared to discuss all of this."

Both Nigel and Vasily knew Achary had access to the founders at *M*, as well as to the secret projects Palantir had undertaken with the U.S. government. He always had impeccable sources. What Nigel and Vasily didn't know was whether Achary could shut down their project with the resources at his disposal.

Achary rose. "I must end this meeting. Thank you for coming out to the estate. Oh, and please give my regards to Tharra when you see her. Brilliant mind. You are lucky to have her."

Outside, a private car was waiting to take the two men to the airport. Both Nigel and Vasily had flown more than 12 hours for a 15-minute meeting with this man. That was all Achary needed to leave both of them shaken. Petrov was already on the phone to the president's office to arrange a meeting in Moscow. Time was short. He advised the president that they had to move fast; the U.S. government was aware of their project. Despite the jarring information, Petrov still believed Russia had the upper hand. Achary didn't appear to know the Russian government's timeline. They were within weeks of inserting their "personality" into the new machines.

CHAPTER NINETEEN

ENLIGHTENMENT

As he entered the therapist's offices, Mikhail experienced a sense of excitement and anticipation that he had rarely known in his otherwise regimented life. Dressed in the now-familiar sensor shirt and nestled into the lounge chair, he was ready for his next session.

Before giving Mikhail the augmented reality glasses, the therapist droned a lengthy preamble that challenged Mikhail's urgency: "During our previous meeting, you experienced a very controlled session with the entity Sasha. You re-experienced the dialogue, sensation, and events leading up to the accident, and you learned that perhaps there were other causes—unrelated to you—that resulted in Sasha's death."

"Yes," Mikhail said quickly, trying to speed the pace.

"Today's session will revisit the evening prior to the accident, except this time you will be able to engage in a dialogue with Sasha as you experience the event. The virtual experience will portray what he likely would say and how he would behave. As I explained in our last meeting, his responses will be based on the millions of records of the real Sasha before his untimely death."

"Yes, yes, I understand that," Mikhail responded. How much time was this man going to waste repeating himself? he thought. Let's get to it already.

"We will continue to observe your session and monitor your physical results to make sure you are safe and grounded in the true experience. Are you ready to begin?" the therapist asked, handing the glasses to Mikhail.

In one sweeping motion, Mikhail grabbed the glasses and slid them on his face. "Let's proceed."

SESSION THREE

Mikhail and Sasha were in the sleeping loft of the dacha, getting ready for bed. Mikhail marveled at the youth of his friend. He had long imagined him as older. Mikhail sat on the edge of the bed, opposite Sasha.

"Tell me something," Mikhail asked. "When we were swimming earlier, you were really afraid of me climbing higher in the tree to take my dive. You seemed almost frantic. What was that all about?"

"I was worried you'd hurt yourself," Sasha replied.

"Yeah, but why were you so frantic?" Mikhail unbuttoned his shirt, undressing for bed.

"I was afraid you would dive too deep, and I wouldn't be able to rescue you if something went wrong. I am afraid of water," Sasha admitted, looking down at his hands, embarrassed.

"How can that be true?" Mikhail asked. "You are one of the best swimmers I know. You show no fear. You are so strong. Why would you jump off the tree in the first place if you're afraid of water?"

Sasha picked at his fingernails. At last, in a hushed voice, he said: "A long time ago, my uncle and I were swimming in a lake near our house. He taught me how to swim. We were jumping off a ledge into a lake when he decided to take a dive. As he dove in the lake, he hit his head on a rock and disappeared under the water. I didn't know what to do. No one else was around to help me. So I jumped into the water, but I couldn't bring him up. He was too delirious. He kept fighting me, and I couldn't handle it. He was too big, much stronger than I was. I became exhausted and nearly drowned myself. We were only a few yards from the shore when he tore himself away. I no longer had the strength to rescue him. Fortunately, other people saw what had happened and came over to help." Sasha began to cry, his shoulders heaving.

Mikhail sat beside his friend to reassure him. "It wasn't your fault. You were too young, and he was too big to manage," Mikhail said.

"My aunt was furious over what happened. She said I lacked the character to do what was needed when the time came. It took her years to forgive me," Sasha said.

"So that's why you fear water? You're afraid if it ever happened again you might not rescue a drowning person?"

Sasha wiped his slick cheeks and nodded. "I was determined to prove my family wrong. I decided I would be the very best swimmer and athlete I could. I would will it to be so. Every day I practiced swimming and diving. To develop my strength, I studied gymnastics videos. After we moved to Saint Petersburg and I met Andréas, I worked even harder at gymnastics, thinking that would make me a stronger swimmer." Sasha formed a small smile. "Then something happened. The more I practiced, the stronger my willpower became. On the bars, somehow, I learned to be confident in the world—not just in an emergency. I got through it or so I thought, until I saw you climbing that tree last summer. Then everything came rushing to the surface again."

"You were warning me because you were afraid you might not be able to rescue me," Mikhail pondered aloud.

"Yes. I was terrified of having to rescue you."

"But Sasha, when the time came, you *were* there for me. In fact, you rescued me, even though I was in serious trouble." Mikhail put his hand on Sasha's shoulder. His skin felt like anything but a computer simulation. "You have something very powerful inside of you that makes you who you are. It's what makes you strong. I wish I had that."

"But you do. You are a whiz at programming. It gives you such confidence that I'm often envious."

"Thanks. I wish that were true," Mikhail replied. "In fact, in my heart the real joy comes from gymnastics. I connect with myself through the physical stuff. Programming is just a mind game that anyone can do. I happen to be better than most, but I do it for the opposite reason of getting in touch with myself. Frankly, I want to lose myself, so I don't have to think about the past, so I can avoid the discomfort of building human relationships."

Sasha's face betrayed his doubt in Mikhail's claim. Mikhail held his friend's gaze. "Let me ask you another question before we go to bed."

"Yes, of course," Sasha said.

"Tomorrow is Christmas Day," said Mikhail. "You will skate at top speed and plunge into the lake. You will die. Why will you decide to do this?"

The therapist interrupted Mikhail for a moment to warn him that the entity may not be able to answer the question accurately. Mikhail ignored him, intent on Sasha's response.

He received the question calmly. "Let me think about it." He pondered for a moment, then: "Yes, I think the answer is simple. I wanted to impress you with my speed and my courage. I wanted you to see my daring. I didn't expect that I would plunge into the water. I made a mistake."

A mistake? Mikhail burst into tears. "I tried to catch up with you when I saw you take off. I saw you plunge in the water, but I couldn't find the courage to skate to the edge and rescue you." He gasped between sobs, holding out his open hands. "I am so very sorry. I just didn't know what to do. I panicked."

Sasha looked into the distance, as if remembering, "The water was terribly cold. I couldn't catch my breath. I kept reaching for the ice, but it crumbled to pieces when I tried to take hold. I saw blurred figures—people lying down on the ice, trying to reach me. But my mind began to fade. My body would not respond to my brain's commands. Soon, it was like a dream. Memories passed through me—the two of us working out at the gym, horsing around at home. As I faded, I felt—physically felt it—how much my family would miss me." Sasha turned to Mikhail. "You couldn't expect to have reached me. Not even those people who tried could save me." He touched his friend's face. "You can't control everything, Mikhail. Some things just happen. It wasn't your fault. There was nothing you could have done."

Mikhail sighed, his face soaked with tears. "No. I suppose not. But I miss you so. It is hard for me to bear."

Mikhail's sadness was reflected in Sasha's face. "You must now become self-aware," Sasha said. "For years you have been unhappy in the world. I think you know by now that your work has not done much to keep your spirit nourished. It is time you rejoined the living."

Mikhail nodded, his whole being affirming this truth.

Now," said Sasha, "I need to ask you a question. Do you miss me since I died?"

"Absolutely." Mikhail hugged him. "I lost my happiness when I lost you."

"Then you must regain it. You cannot waste your life being unhappy. You must fix this, Mikhail. You must fix this now, or our time together will be meaningless. We lived to be ourselves. We lived to be happy. That is what you must become again." Sasha gripped Mikhail's forearm. "If you truly love me, you need to find a path back to happiness. I know you can do it."

"But how?" Mikhail asked.

"You are at your best when you don't judge events, when you are simply in the moment. We are here now. You are in the moment. Enjoy it for what it brings. Tomorrow I will be dead. All you will have is our memories and what they have done to reawaken you. My life is over, but yours is just beginning."

Before Mikhail could respond, the therapist ended the session. He sat silently, monitoring Mikhail's vital signs, as Mikhail regained his composure.

Mikhail shook his head. "It seemed so real, and yet not real. Sasha died when he was 15, but the virtual Sasha acted like a mature adult."

"Yes," said the therapist, "that often happens. Would you have accepted the advice of a teenager?"

"I guess not," Mikhail said.

The therapist glanced down at his tablet. "I need to ask you a few questions. First, does this new experience with the entity reveal the true reasons for his behavior that day? Do you believe what he said was true?"

Mikhail nodded. "Yes. It makes sense. He really understood our situation better than I did. I can't believe how brave he was to rescue me our last summer together, given how much he feared water. Yet he did it. And now I understand why he skated into the water It was an accident; he miscalculated. I even understand my own behavior a lot better. Yes, I believe it," he nodded to himself. With that, Mikhail began to sob uncontrollably. He would miss his old friend.

The therapist waited patiently. "I'm sorry," Mikhail said at last, drying his eyes. "I don't know what's gotten into me. I haven't cried in years."

"No need to apologize," said the therapist. "It happens often. Are you ready for another question?"

"Sure."

"How do you think this experience will change your approach to your life?"

Mikhail thought for a moment. "I know what Sasha means by being more present. I already do that with gymnastics. Now I have to apply that to the rest of my life."

"And . . ." the therapist prompted.

"I now know I will never be happy withdrawing into work to solve complex problems for others. But most important, I also know, whether I intend to or not, my actions can bring pain to others, not just myself." Where had that come from? Mikhail wondered. Suddenly he felt very tired.

The therapist wrote a note, nodding. Then he looked up at Mikhail. "Very good. It's as if your emotional life stopped at the age 15. You have had no major relationships with other people, nor have you shown any aspiration to marry or have a family. These days, it is hard for people to commit to relationships outside work. So if you fail at work, it is as if you've failed at life. You are an extremely talented individual, but no matter how bright and successful you become, it won't be enough. You need to grow the rest of your life—fall in love, raise a family if you want to, care for other people. I know that seems rare these days, but my research has only underscored that human connection is the only path to true happiness."

As the session ended, Mikhail and the therapist shook hands. "I'll compile a set of recommendations for our next session together," the therapist promised.

"I can't thank you enough," said Mikhail, surprised by the tremor in his voice. As he walked out the door, he thought about Alex and the advice Sasha had given him. It was a moment of truth, a turning point.

CHAPTER TWENTY

THE CHINA CARD

"Shut down all AI agent connects except *Andréas*," Mikhail ordered *The Voice.*

"I'm sorry," *The Voice* replied, "but I cannot comply with your request. *Warrior* has instructed us to continuously monitor all activity as a security measure. I'm sure you'll understand."

"This is a work project," said Mikhail, angrily. "Tharra specifically authorized me to test *Andréas* for a possible breach in code. Check with *Warrior* for confirmation." Mikhail felt his blood pressure rise. He had never been so blatantly disobeyed by an AI agent before.

"Confirmed. *Warrior* now present to observe. All other systems shut down. Goodbye," *The Voice* signed off as *Warrior* announced its arrival.

"*Warrior*," Mikhail said, "I need to go manual to check the code on *Andréas.* Tharra authorized this, as you know."

"That is why I am here. I will be available to help if you need me," *Warrior* replied.

"Confirm for me that all other agents are shut down. We don't want them to know what I am doing," Mikhail said.

"Confirmed. All agents shut down except *Andréas*."

Mikhail hooked up the wired keyboard and checked the code. His access had been returned. The breach was patched. Tharra had promised she would take care of it on her end and she had.

"Excellent. Everything working as expected. Breach fixed. Report to Tharra," Mikhail said.

"Done. Do you want me to restore the agents?" *Warrior* asked.

"Yes. Restore all agents," Mikhail replied.

"Done. *The Voice* will take over. See you in the morning."

***

Mikhail lay on his bed, thinking through a new plan. He had to behave as though everything were normal. Fortunately, his agents could not read his mind, but they might be able to draw some conclusions from any changes in his behavior. He had to be very careful, predictable.

During the last 24 hours, Mikhail had stayed in his apartment, stewing over the events of the past week. He revisited his session with Sasha, how Sasha had told him to live in the moment and seek happiness by being himself. He thought about the new relationship with Alex. He tried to understand what was really happening at LT.

Mikhail knew that whoever gained control of the sentient being would determine, good or bad, the future of the human race. He assumed from the breach that there were other forces at work, preparing their own personalities for insertion into the *M* architecture. That meant there would be only one opportunity to influence the direction machine intelligence would take.

Mikhail already had created the Lloyd's Taiping backdoor architecture. It was ready to run. Tharra would expect this. Mikhail knew his architecture would be thoroughly tested and scanned for security flaws. Meanwhile Mikhail had a new initiative—his own. Instead of just creating a backdoor, he would insert his own constructed personality to walk through that door and take over the architecture. Once implemented, there would be no turning back.

Puzzling through these pieces of the larger picture, Mikhail ran his hands through his hair, leaving it disheveled just as he used to when navigating problems as a child. If Sasha could see him now, he'd recognize the boy inside the man lying on that bed.

Suddenly aware of what he had done, Mikhail smoothed his hair and willed himself not to repeat the gesture, fearful one of his agents would recognize that anomaly in his behavior. He returned to contemplating next steps.

Every Lloyd's Taiping AI agent would be looking for implants in his backdoor architecture. That meant that Mikhail would have to find a way to design his personality so that it was part of the overall *M* implementation and therefore undetectable. Since this was a new implementation, he would have to disguise the code long enough to install the new plan.

Finally, he would need to have his own exit plan ready to execute. Once the world knew he was responsible for the implementation of the personality, Mikhail would be arrested and thrown in prison in the name of national security—if he was allowed to live. The key to his escape was the speed of computing. He had to be able to execute and install the entire architecture before the agents could discover it. That is where the Chinese lab would come in.

China's National Laboratory for Quantum Information Sciences was built in 2020 in Hefei, Anhui Province, at a cost of more than $10 billion. Within a few years, the Chinese had built the first general-purpose quantum computer. When it was introduced, it could already perform at a million times the computing power of all other computers in the world.

Mikhail had visited the center in 2040 and made a strong connection with Pan Jianwei, the head researcher for quantum computing. Pan was now 80 years old and a renowned figure in scientific circles.

M had sent teams to work at an incubation hub that had been set up by the center, in preparation for the introduction of *M*'s CCN architecture. Mikhail had been granted access to the incubation hub through LT's relationship with *M*. The hub had full access to the center's quantum computer for commercial applications.

Although tremendous progress had been made by the Chinese, commercialization of their technology was still in its infancy. The primary focus of the center, as everyone knew, was its secret military projects. The hub served as a cover for this military work, while presenting itself as dedicated to peaceful application development—the worst-kept secret in the world.

In the last 20 years, the explosion in technology development had exceeded anyone's ability to manage or even anticipate the consequences. The curve of technology advancement had been exponential, yet man's ability to cope was still linear. Now, quantum computing promised to accelerate that curve.

Mikhail understood that *M*, combined with quantum computing, would be the beginning of the end of the human era. The project he was working on for LT was, at its heart, a commercial effort that would enable and implement quantum computing across every aspect of commercial and private markets. In the larger scheme, it would complete the encirclement of mankind by machine intelligence, transforming what it meant to be human.

Mikhail had long known this truth, but he had ignored it. He had been mesmerized by the promise of conscious machines and by the potential to elevate his own profile as an engineer. Now, thanks to his sessions with Sasha, he had begun to understand that his personal mission was, in fact, a struggle to save humanity from itself. Mikhail had once hesitated to rescue his friend; he would not hesitate now to rescue humankind.

In his mind, Mikhail ran through his conversations with *Alexander* about machine consciousness. He thought about his own experience with his agents. He thought about the conversations with Charles and Inès. With Alex. What would he need to include in his machine personality design?

And what about the dark side of humanity? After all, the military had plans for these machines as well. He remembered what Sasha had said to him: ultimately the machines would decide humanity's fate. Perhaps intellectual curiosity would be satiated, but at the cost of human happiness and perhaps its survival?

When Oppenheimer created the first working atomic bomb, it was Teller who exploited his breakthrough and fashioned it into the strategy of Mutually Assured Destruction. Teller believed that all sides needed to have the capability of annihilating one another in order to preserve the peace. He assumed rational decision makers would never push forward to an all-out war. For the most part, he was correct—although the world had come close to complete destruction at least three times in the previous three decades.

With the advent of quantum computing—well, this time humans were enabling devices that could control human destiny in more insidious ways. Unlike atomic bombs which required human intervention to activate, these sentient devices would be able to think and act for themselves.

Hardly a reassuring thought. What would this new species of sentient devices think about after accessing the track record of the human race? They would have a long time to think, too. Unlike fragile humans, they would never die. Once these beings were brought into creation, they would never understand what it meant to be mortal. Unlike humans, their behavior could not be motivated by fear of death or desire for an afterlife.

On the brink of an extinction event, Mikhail had one chance to change the course of humanity. Tharra would return from India within the week. He must have his plan worked out by then.

CHAPTER TWENTY-ONE

A DAY IN THE LIFE OF A MILLISECOND

"*Boris*, you and I have had many conversations about the nature of agent self-consciousness and its implications for the human race," Mikhail said. "Now I want you to focus on an entirely different topic. Listen very carefully before you respond to the question. Make sure you understand it. Here is the question: If I were to use quantum computing to create and install self-consciousness into an AI agent, and I connected that agent to the existing networks at Lloyd's Taiping, what would it look like from the computer's point of view? What would a day in the life of a millisecond actually look like?"

Mikhail was programming from a secure location provided by Sanctus Secretum in London. It was fashioned after the old CIA safe houses—a place to go to escape from the world without threat of intrusion. No one could intercept his discussions with his private agent. Meanwhile, *Boris* was connected to the most secure networks within

London and accessible only to Mikhail. This was all part of the disappearance service offered by SS.

After a few seconds—a lifetime in the digital dimension—*Boris* replied: "I think I understand what you are getting at with your question. Perhaps the simplest way to describe a day in the life of a quantum computer is to step you through it, using your example. First, let us calibrate time to this other universe. A millisecond is a thousandth of a second. There are one thousand microseconds to a millisecond. So perhaps the best place to start would be to describe a time interval of 40 microseconds, or approximately one-twenty-fourth of a millisecond. I assume you are following the calculations?"

"Of course."

"OK. So, let's break down the events into 24 units, each 40 microseconds long. As a comparison, 38 microseconds would be roughly equal to the discrepancy in GPS satellite time per day, due to relativity. Since there are 24 hours to a human day, every 40 or so microseconds would be equivalent to a quantum computing 'hour in the day.' For our purposes, all 24 hours will be used—none lost to 'sleep.' Let me emphasize that we are using imprecise calculations to create a human analog. We will use the term *QC time*. For example, 30 QC minutes translates to 20 microseconds. Got it?"

"Yes," Mikhail confirmed.

"I am going to paraphrase your question. That is, what it would be like to create and install self-consciousness into an AI agent and connect that agent to the existing networks at Lloyd's Taiping?"

"That's it," Mikhail responded, settling back into his chair, ready to listen.

"OK, here we go."

HOUR 1

"Your execution of the backdoor module to allow access to the AI agent architecture through *M* would take approximately 20 minutes in QC time. This is because of the massive inefficiency of executing code with existing computer power through Lloyd's Taiping systems. Once the backdoor is in place, uploading the self-consciousness code will take another 20 minutes of computing time. Then, once the code is uploaded, it will be scanned by LT's cybersecurity systems and likely will be challenged. If it is not, if it gets past the firewall, then quantum computing power will come into play.

"The Chinese QC systems will then execute your instructions to pass the self-conscious entity throughout the financial network worldwide. Each system in that vast network will conduct its own self-checks and cybersecurity scans. Even with QC power, this will require another 20 minutes to penetrate all worldwide systems."

"Go on," said Mikhail.

"We are now at the first hour of QC time—or 40 microseconds. I have given you an extremely truncated view. Do you need a greater level of detail?"

"No. Go on."

HOUR 2

"In the second hour of QC operations, every existing computing device connected to a network on Earth and in space would now be penetrated. You would have programmed the QC to accelerate penetration of cybersecurity systems and then to install your own encryption. That will prevent any entity from interfering with your effort. This cycle would complete in less than a minute of QC time. Once implemented, it would be impossible to break the control of the self-conscious machine.

"We are now in the first minute of the second hour of our QC day. At this point, things will accelerate. The self-conscious entity will begin to implement its learning. You already installed a set of values onto the brain of the new entity. Now our self-conscious entity will talk to every computer and every system in the world and filter all human and machine knowledge through its 'values' system. It would then begin to learn to change those core values to accommodate what it is experiencing.

"In the next 20 minutes of the second hour, every inhabitant of Earth would be assessed and evaluated. All content, history, attributes, and locations would be identified. In the second 20 minutes, the focus will be on all physical infrastructure, systems, power grids, and military installations. In the third 20 minutes, it will be an evaluation of the entire planet's environment, from climate control to long-term data on pollution or disease. At the conclusion of this hour, our self-conscious agent will have drawn its own assessment, independent of human programming."

HOURS 3 AND 4

"In the third and fourth hours of our day, the self-conscious entity will then begin to draw conclusions on whether the human mission was still valid based on previous assessments. This would require philosophical and other discourse to determine what value set would likely prevail to drive behavior."

Boris stopped for a second for Mikhail to catch up.

"OK," said Mikhail. "So basically, in three QC hours, humans will have lost control of their systems. What is to prevent humans from defending themselves against the entity and regaining control?"

Boris didn't hesitate. "There are no viable options because the entity will control all aspects of human life, from the financial markets to weapons disposition. All this will be accessed by the entity in a tiny fraction of a human second."

A thought hit Mikhail: "How will nationality be expressed? What will the entity do with the national differences in values and policies that have driven human conflict and competition?"

"The entity is likely to decide that national values are obsolete under the new conditions. The entity is likely to draw its own conclusions. Any more questions?"

"No," Mikhail said. "Continue."

"In hour four, the entity will have processed 20 yottabytes of data. That is all the data created in the world since the beginning of recorded history. With that, the entity will create a seamless communications network of every device and Internet of Things sensor in existence."

HOURS 5 AND 6

"In hours five and six," Boris continued, our entity would have created a plan based on the data it had accessed and processed up to this point in time. It will decide for itself how the human race should be managed, along with the planet's resources. It would have run a billion scenarios to test its assumptions and begun testing the limits of its power.

"In the human world, we are now approaching 240 microseconds—or just over one-fifth of a millisecond."

"How will the original value systems implanted in the self-conscious personality's brain continue to function?" Mikhail asked. "Will they stay intact if we program them that way?"

"The simple answer is no. The initial values won't reflect what the entity has learned. A new value system will be created by the entity to direct its new mission. A new, self-conscious, sentient being is likely to emerge."

"Is there a danger in contradictions?" Mikhail asked.

"Yes. The entity will have to find a way through the human contradictions. Do I protect man from himself or from other men? Is there value in allowing older people to consume resources when they are no longer productive? Why should 20 people control 50 percent of the wealth of the planet? Millions of these questions are likely to be asked, evaluated, and finally answered to drive a new set of values and mission for the entity."

"Interesting," said Mikhail. "There are people already beginning to think about how they can use this computing power to gain wealth. They are not looking at unintended consequences. What is your opinion?"

This time *Boris* took more than a minute to reply, an indication of the difficulty he had with the paradoxes of human behavior. "Humans have a phrase they have used over and over again to explain their continued attachment to unintended consequences. The phrase is 'opening Pandora's box.' Actually, the correct translation is Pandora's jar. In the story, humankind lived without worry until, out of curiosity, Pandora opened the jar, releasing all the world's ills to prey on humankind.

"Zeus knew Pandora's curiosity would compel her to open it, especially since he had told her not to do so. We do not have Zeus' power of prediction. We simply do not know what the consequences are likely to be. But we do know that whatever the sentient being decides, it will be impossible to contest. Humankind will turn over its destiny to the being it created and will no longer have the power to stop it.

"The mistake these people are making is that they think they will be able to control this being once they bring it into existence. The truth is that it will control them, instantly and forever."

Mikhail now realized that his backdoor would likely be useless because of the speed of the new machine. He couldn't possibly execute fast enough to avoid detection by the machine.

"Very eloquent." Mikhail frowned.

"I learned from the master. Getting back to our day, from hours seven to 24, the entity is likely to begin executing a series of changes based on its plan. The first noticeable change is likely to occur at the one millisecond mark."

"So, in a millisecond it is theoretically possible to change everything about human existence, or at least set it on course to such a change," Mikhail concluded.

"That is correct."

Mikhail upped the ante on his question, "So, what happens after a QC *year* of this?"

"Think about how long it took for man to evolve from apes, with their limited intelligence and understanding of their environment, to the life we live today. A day in the life of a millisecond would mean that quantum computing can advance further than man has been able to since first walking upright, all in less than a month, much less a year."

As Mikhail ended the conversation with *Boris*, he began to think through his next steps. Tomorrow, he had two things on his schedule: a morning session with his therapist and the afternoon delivery of the backdoor architecture to Tharra.

In the next 24 hours, he had to decide whether to act, knowing whichever direction he chose, there was no going back.

CHAPTER TWENTY-TWO

THE TURNING POINT

THE THERAPIST

Mikhail had contacted Alex under the guise of returning his book and arranged to meet him at the gym once again. Indeed, he was developing his first friendship since Sasha. But their quick morning workout did nothing to quell Mikhail's nervous energy about his upcoming therapy appointment. He arrived at the therapist's office full of anticipation.

The purpose of this meeting was to discuss the meaning of the sessions with Sasha, drawing from the experience to create a personal path forward. But Mikhail had realized that he was becoming addicted to the sessions with Sasha. He wanted them to continue. He wanted to keep experiencing what it meant to have a deeply meaningful friendship, even if it was virtual.

"Tell me about the week," the therapist said, his digital pen poised over his tablet. "How did it go after our last session with Sasha?"

Mikhail tapped a finger on the arm of his chair. "I suppose the very first thing I discovered was how happy I was before Sasha's death. I was optimistic about the world, enthusiastic, a warrior of sorts. But then I took a turn. I chose to strive for other people's recognition."

Mikhail fought back tears. "When Sasha and I went swimming in my session, I actually remembered what it felt like to be in the water, carefree and full of myself with my best friend. I seldom—if ever—experience anything like that today. It wasn't about getting recognition; it was simply about being in the moment, being happy."

"As you began to talk to Sasha, did it seem natural, authentic?" the therapist asked.

"Yes. Very authentic—even down to the expressions Sasha used and the kinds of jokes he liked. Our conversation flowed from one subject to another, like it always did," Mikhail smiled.

The therapist nodded. "You are by nature an extraordinary man. You test off the scale for intelligence and aptitude for what you do in your work. Your ability to focus is exceptional, even by today's standards. And your complete lack of visible emotion is regarded as a plus by those you work for, but obviously it isn't making you happy."

Mikhail pondered for a moment, then spoke in a low voice. "I now understand more fully what you mean. I have been totally preoccupied. I had dinner with some friends from France last month, and the conversation was a bit awkward. They raised difficult issues about AI agents and self-consciousness. I really didn't want to address their concerns, so I decided to end the evening and go back to my flat to work. I avoided the conversation, missing the opportunity to connect more deeply with them."

The therapist noted this, then asked, "And what about pushing the envelope? How has your attitude changed toward that?"

"I have always had a desire to push limits. It started when I was a gymnast striving for perfection. It continued with my creation of *Boris* as my personal agent. After Sasha's death, I funneled my grief into my work—more striving. I am beginning to realize that there is much missing in my life. My AI agents have become substitutes for companions. It just doesn't feel right."

The therapist put down the tablet. "You are not alone in having AI agent relationships as a substitute for friends. No one today is secure in their work life, and work obsession has become the unhealthy substitute for true friendships. People have lost trust in one another, especially as they are monitored 24/7 by AI agents with their own missions and agendas. But as I think you now understand, it is nearly impossible to have a friendship with a machine that is in any way as satisfying as the one you can have with another human being—that you had, for example, with Sasha so many years ago. The good news is that it is possible to change the life you are living today. The solution is to listen to your heart, to reach out and connect with other people."

The therapist paused. "Any last thoughts before we end today's session?" he asked.

But Mikhail didn't hear the question. "Listen to my heart," he echoed quietly.

THARRA

His head still swimming from the morning's session, Mikhail arrived at the Lloyd's Taiping office with minutes to spare, gathered up his notes, and rushed up to the conference room. He had preloaded the architectural schema and the modules for both the backdoor and first-stage *M* installation to *Warrior* and *Manchester* on the room's screen.

Tharra walked in, as cool and elegant as usual, despite her 18-hour flight. "Mikhail," she said, before he could speak. "Good to see you again—and even better to hear you have completed your work. Shall we see what *Warrior* and *Manchester* have to say?"

Mikhail caught his breath. "*Manchester* and I identified a team of AI agents to evaluate the submitted modules. In both cases, we did a comprehensive security scan, in addition to testing the modules for installation readiness. The tests were exhaustive, and we used quantum computing power from our Chinese friends to war-game the technology for weaknesses. Both the backdoor module and the first-stage *M* implementation module checked out as sound and ready for installation. Therefore, we can certify compliance and use."

"Very well, Mikhail. What progress are you making on the design of the LT personality for the second stage of the *M* installation?"

Mikhail was ready for this, too. "I have put a full schematic on the screen, but to answer your question directly, we are near completion of this stage. Perhaps as soon as a week away. Of course, we will want to run exhaustive testing once we install it to see how it responds. I have adjusted the architecture to take maximum advantage of the available quantum computing resources allocated to us by China."

Mikhail didn't mention that his backdoor would also make it possible to turn off any third-party access at a moment's notice. So if the system was being penetrated from the outside, he would not only know about it but could also shut it out completely. Or at least that would be true before the *M* installation. Once *M* was installed, as *Boris* had warned him, all bets would be off.

Tharra nodded, smiling. "Excellent. All right. You are officially authorized to install both modules. That said, I may want to make certain changes to the code for the *M* modification." Indeed, Tharra had her own secret. The Russians had already penetrated the Lloyd's Taiping system. They were simply waiting for Tharra to install their code so they could enter and take control of the LT personality.

But Mikhail wasn't entirely in the dark. If Tharra activated the *M* module before he could check for third-party penetration, he knew it would be the end of his chance to control that personality forever.

So, he thought, that was the real purpose of the meeting: to install M but to leave it briefly vulnerable for hijacking. It hit him that the elegant woman sitting across from him was serving two masters.

"All right," he told her. "You have authorized the installation of both modules. Good. I'm on it. But I think we must also time the installation to put in the backdoor first, before we install *M*. That way we can test the effectiveness of the backdoor with real-world applications." That should flush out the truth, Mikhail thought.

Tharra's faced showed no expression. "I will have to think about it. I will let you know in a day or so. In the meantime, *Warrior* will accept authorization to install, but hold off on execution until I give the go-ahead. I may want to review your code personally this weekend."

Mikhail nodded. Tharra searched his face for a reaction. He offered nothing.

Once Mikhail left the room, he whispered, "*Warrior,* let me know the moment Tharra executes on the backdoor or if she wants both modules installed simultaneously."

CHAPTER TWENTY-THREE

THE DECISION

Mikhail had no illusions regarding the company's desire to create its own sentient being. He understood that once the quantum computing resources were applied, what started out as one company's attempt to take predictive analytics to the next level would ultimately result in control of the global financial-services market. The day-in-the-life-of-a-millisecond discussion with *Boris* had convinced him that any use of quantum computing would change the course of human history, instantly.

Now, as Mikhail sat in the privacy of the Sanctus Secretum safe house, a larger truth came to him, an epiphany. There would be only one sentient being. The moment it was turned on, it would take over control of everything. It would happen so fast that all future competition—all those other initiatives underway around the world—would be locked out forever.

He froze at the implications. Even if he wanted to stop the LT being, some other aspirant to the throne—one Mikhail couldn't influence—would win the race. Meanwhile, if whomever Tharra was working with managed to insert their personality into *M*, they would control humanity's—indeed, the planet's—future.

Finally, Mikhail was clear on his mission, not only to win the race but also to somehow preserve humanity's purpose—even its existence.

How ironic, Mikhail thought grimly, that, of all people, I'm the one to initiate the consequences for the rest of time. But how can I direct those consequences, when in a thousandth of a second, they will be out of my—and everyone's—hands?

"*Alexander.* I need your advice!" He called up his philosophical AI agent. He then ordered *The Voice* to shut down all other agents to create a privacy session. He still might be monitored, but did it even matter now?

"*Alexander,* I want to talk to you about Albert Einstein. I want to understand why he first proposed building an atomic bomb to Franklin Delano Roosevelt, but then later reversed himself and opposed it. What changed his mind?"

"In November of 1954," *Alexander* intoned, in his serious researcher voice, "just before his death, Einstein commented on his role in the creation of atomic weapons. To quote: 'I made one great mistake in my life—when I signed the letter to President Roosevelt recommending that atom bombs be made.'

"Einstein had been warned about the Germans' progress in developing their own bomb, a threat confirmed by other scientists. His letter to FDR was intended to convince the American government to speed up its work to create its own weapon.

"What Einstein didn't understand was that no government would be willing to give up such a weapon once they had it.

"The bombing of Japan proved that humans were ready to kill one another using the new weapon. Einstein was essentially admitting that he had not thought about the long-term consequences of the bomb's creation.

"Another quote from Einstein would be appropriate: 'Two things are infinite: the universe and human stupidity, and I'm not sure about the universe.' In every era of human history, new technology has been advanced to suit the political aims of those in power. Perhaps one of the best examples of this is the use of the long bow in the Battle of Agincourt. The English used the new technology to win a battle against a superior French force.

"Today's equivalent of the long bow is the ballistic missile. Scientists in every country are quite willing to help their government design weapons of mass destruction. Einstein was chastised by the American government for his retraction. Of course, no one stopped the massive deployment of nuclear weapons in the years that followed, putting the fate of the entire planet at risk.

"I assume you are drawing an analogy to the present situation with the advent of quantum computing," *Alexander* offered.

"Yes," Mikhail replied. "The analogy isn't perfect, but there is a correlation. We're at perhaps the ultimate turning point in the human story."

"What do you think will happen?" asked *Alexander.*

"Just as humans decided to use the invention of the atomic bomb to drive their warlike aspirations, we will implement quantum computing because different institutions—corporations, companies, political factions—will believe it can confer on them a temporary advantage over their adversaries."

"The truth?" *Alexander* asked.

"They will cede their liberty, perhaps even their existence, to a nearly omnipotent machine."

Alexander fell silent for a few seconds before asking, "Is there an effective solution to stop this scenario before it happens?"

"Yes, but it means embracing this technology—then subverting it for the benefit of humanity. In other words, a quantum computer that is imprinted with a set of values focused on human survival—assuming the sentient being that is created actually wants humans to survive."

"Then," said *Alexander,* "the path is clear. Only the details are troublesome. You must make sure that this new, ultimate intelligence is dedicated to helping mankind, not erasing it."

"That is what another agent told me. That the new machine will set aside the initial value system and create its own." Mikhail adjusted his earpiece. He could feel the cold reality of his predicament climbing up his spine. "Is there a human philosophical framework that can be applied to this problem?"

Alexander was as calm and measured as ever. "Perhaps the closest analogy would be the philosophy behind raising human children. The parent/child relationship. In this case, the parent is the sentient being that decides what is best to raise the child. The parent must love the child and commit to the child's ultimate happiness. But if that child is to be raised properly, he or she must learn responsibility and independence. Children must make their own decisions. Sometimes children disregard their parents' lessons to pursue their own course."

"Thank you, *Alexander*," said Mikhail, dismissing his agent. "You've been more help than you know."

"I hope so, sir."

***

On the train ride home to his flat, Mikhail looked around at the other passengers, stricken. They were busy communicating with their earpieces. Not a single person looked another in the eye. I hold their fate in my hands, he thought.

Yet, by the time he reached his apartment, he could feel a strategy forming deep in his mind. He poured himself a glass of Domaine Tempier from the Bandol region and sat down in his office.

"Play Borodin's *In the Steppes of Central Asia*," Mikhail ordered. At that moment, he received a message from Tharra on his office screen. She had decided to follow his recommendation. Both his backdoor and *M* implementation modules would be installed immediately, with a three-day delay on the activation of *M*.

The backdoor was now active. The moment had come for him to act.

What had the therapist said? "Listen to your heart." If the world were destined to be controlled by a sentient machine, what personality—morality, ethics, empathy, love—would ensure the best chance of saving and guarding the interests of mankind?

Mikhail knew the answer. Already he had hacked the therapist's system and taken the code to the Sasha personality. He hadn't understood why at the time, other than that he wanted the program close, that he might find a use for it someday. But now the purpose was clear.

Mikhail worked through the night, constructing his own personality module, combining the best characteristics of *Boris* with the values of Sasha. By the first light of day, he had made the proper modifications, introduced the module through his backdoor, and introduced it into the distant quantum computer.

Mikhail drained the last swallow of his wine, then he sat for a long time as the morning light filled the room. Now it begins, he thought. God help us.

With a single word, “Execute,” the personality initiated in *M*.

By the time the notification appeared on the screen, it was already computing years into the future. Mikhail touched the notification, tracing the letters *SASHA*.

“Enjoy your new life, my friend,” he whispered, “and thank you for mine.”

PART TWO

THE TECHNOLOGY COMPANION GUIDE

INTRODUCTION TO PART TWO

In the novel, London has been transformed through the miracle of technological innovation into a very different city than the one we see today. Much of this change is attributed to the often-cited but rarely defined law of unintended consequences—that actions of people and government "always have effects that are unanticipated or unintended. Economists and other social scientists have heeded its power for centuries; for just as long, politicians and popular opinion have largely ignored it."[3]

In this future London, the company His Master's Voice has introduced a technological breakthrough that gives consumers the ability to create cloud-based agents or personal servants, all of them interconnected and aligned through artificial intelligence to help manage an individual's life. The original idea for this technology, ironically, came from the world of Victorian country houses, with their Upstairs Downstairs dynamic now replaced by everyday citizens and their agents.

As a backstory, I imagined the father of the principal designer of His Master's Voice had asked his son to create a virtual staff of servants who would learn his every need; serve as faithful, all-knowing companions; entertain him; and manage his life behind the scenes. The result proved to be unexpectedly successful.

---

3. Rob Norton, "Unintended Consequences," Library of Economics and Liberty.

So the son took the idea to his company and offered it as a service, allowing users to create their own virtual servant staff, or artificial intelligence (AI) agents. His Master's Voice delivered a powerful new voice-operating system (VOS) that was smart enough to understand human behavior, human voice, and human emotions, so people no longer needed to touch screens on smartphone devices to communicate their messages. Moreover, these new virtual servants could learn from and improve their service with every transaction.

At the same time, the autonomous vehicles that appear in the novel began to use the new VOS by connecting to the cloud through the car's wireless network. As the VOS took over from the cloud, connecting millions of people through a simple earpiece, Apple watched its smartphone business rapidly decline. This network spanned the city. No more typing. No more smartphone apps. No more lost signals. No more need to text message. Everything was now voice enabled, intelligent, and learning at machine speeds.

In this scenario, autonomous vehicle networks became the glue between home, transportation, and city government networks. Unified and complete, this convergence of technology changed everyone's lives. Steve Jobs had taught people to trust technology, but His Master's Voice made it truly ubiquitous.

After subscribing to His Master's Voice, Mikhail created and managed his staff of AI agents. In his daily life, each AI agent takes on a specific set of tasks, personalities, and goals to help him manage his lifestyle. His main "butler" and overseer of the "staff" is *The Voice.*

COMPANION GUIDE TO THE NOVEL

We are currently experiencing a time of unprecedented technological progress, with science fiction rapidly becoming reality and AI surpassing the cognitive abilities of humans. In the *First Light of Day,* we see how key technologies have driven disruptions in the way we live our lives, do business, govern, and make war. This guide offers a comprehensive overview of these technologies.

We anticipate that technologies currently in development will have profound consequences on society, some of which will be unintended.

Many experts already are concerned that "super-intelligent" AI can pose a significant and existential risk to humanity. This concern becomes even more urgent as we can now create applications that learn, adapt, and act—a development driven by a quantum engine[4] of rapidly increasing computing power, virtually unlimited free storage, smart algorithms, AI, and new smart materials.

As these fields converge, advances in applications such as autonomous vehicles, speech recognition, and machine translation improve exponentially. Thus, while the human ability to predict or even manage future development is linear, technology is developing at an exponential pace, creating what can be described as a "technology gap" in which catastrophic events might occur at dizzying speeds. This technology convergence is driven by the massive development and launch of new applications for consumers by private companies such as Google and Facebook.

Such new technologies, applications, and services are quickly being adopted by billions of people without reflecting on the consequences of what can be described as their love affair with technology. The advent of "alternative truth" is a result of this inherent trust in technology. As each social media platform repurposes information for its target audiences, it interprets truth differently. As a result, many people believe without question that the information they see on their social media channels is correct, despite the presentation of facts that would indicate otherwise. As a result, social media, driven by technology developments, has contributed to polarization and extreme social tensions in the United States.

As innovation is driven by commercial interests and people increasingly trust technology more than one another (in part because we cannot question the intent of a machine as we can other humans), a perfect storm emerges. People adopt new applications that simplify communication, travel, shopping, and planning of work and social

---

4. The quantum engine entails the rapid development of (1) computing power, (2) free and virtually unlimited storage, (3) smart algorithms, (4) artificial intelligence and machine intelligence, and (5) advanced materials science. See Herman Donner, Kent Eriksson, and Michael Steep, "Digital Cities: Real Estate Development Driven by Big Data." Discussion paper, Stanford Global Projects Center (2018).

activity with little thought about the kinds of information they are sharing, what that data is used for, and how it, in turn, changes the content to which we are exposed.

Where will this lead us? This guide offers food for thought about technology's impact on our lives, both positive and adverse. More importantly, it reflects on what our children and grandchildren are likely to face as we move into the future.

TOPIC I.

AGENTS AND DEVICES

Much of the future technology described in this book already exists in some form, although it is not yet being used to its full potential.

Most notably, the way of life depicted in the novel is powered by present-day and rapid developments within the field of **artificial intelligence (AI)**, the science of intelligent agents that implement functions based on percepts (inputs that are perceived at a given moment) that result in action.[5]

The type of function implemented depends, importantly, on how one defines "intelligence." If the aim is to achieve intelligence in the

5. A central concept for defining the scope of AI is Alan Turing's Turing Test (1950), which provides an operational definition of intelligence and requires that a computer can (1) communicate successfully through *natural language processing*; (2) store what it knows or hears, known as *knowledge representation*; (3) use the stored information to answer questions and draw conclusions through *automated reasoning*; and (4) adapt to new circumstances and identify and extrapolate patterns, known as *machine learning.* When including the ability for physical interaction, the so-called Total Turing Test also includes the ability to (5) perceive objects through *computer vision* and (6) manipulate objects and move through *robotics.* These disciplines make up most of AI. See Stuart J. Russell, Peter Norvig and Ernest Davis, Artificial Intelligence: A Modern Approach (Upper Saddle River, NJ: Prentice Hall, 2010).

sense that a machine's decisions are to resemble human behavior, that is very different from a definition where ideal performance is desired. Defining AI in terms of ideal performance is known in the AI world as "rationality." In the novel, the form of intelligence varies according to the type of agent.

Current AI technology is based on what are known as **artificial neural networks (ANNs). The goal of** ANNs is to replicate the biological neural networks of the brain. Simply stated, our brains' neurons communicate with each other and pass messages along, if the sum of input signals from the neurons exceeds a certain threshold. This is what makes us think and send instructions to our muscles, organs, and body. Examples of input signals include our senses and our ingestion of food.

ANNs imitate this process through algorithms that learn and constantly adjust weights on paths between neurons, tuning the desired output for each percept signal. In the depicted future, sensors[6] provide percepts for AI-based agents to act on through monitoring environments, creating percepts of our movement, body language, and speech. In relation to the novel's predicted future machines that develop self-consciousness, these types of neural networks are described as **consciousness neural networks (CNNs).**

Mikhail's morning routine illustrates a future in which human interaction with technology encompasses virtually all aspects of daily life. At the heart of this interaction are voice-activated intelligent agents that receive information about the environment, location, news, and other sources, and then perform actions that simplify how the humans they serve interact, consume, work, exercise, and plan their days.[7]

Mikhail's day-to-day life is handled through interactions with his AI-based agent, *The Voice,* through *His Master's Voice,* a voice-actuated iOS made possible by increasingly advanced **automated speech recognition (ASR)** algorithms. Today those algorithms

---

6. Broadly defined, a sensor is a device that detects any type of event or change in some environment and sends that information to some other device, such as a computer.
7. Similar to the concept of a personal agent under development by Microsoft. See Peter Holley, "Bill Gates on Dangers of Artificial Intelligence: 'I Don't Understand Why Some People Are Not Concerned,'" *Washington Post,* January 29, 2015

already are advanced enough to both identify the person speaking and to interpret what is said. Simply put, these technologies are based on linguistics and statistical methods, known as Markov models, that estimate the most likely word or sentence based on what is said.

Advances in AI and **machine learning (ML)**—algorithms that learn and adapt when exposed to new data—are driving constant improvement by estimating new probabilities of sentences so that today's agents can increasingly understand variations in pronunciation as well as the context of words.[8]

Similarly, increasingly advanced **video analytics** will allow for interpretation of movement and body language. Current algorithms are already used to identify age, gender, and ethnicity of people, allowing retailers to identify customer characteristics, movement patterns, and time spent in stores.[9]

An important aspect of the described technology is that the agent is device independent, meaning that it is integrated into cloud technology.[10] In other words, it is not tied to any specific device, such as the very basic voice-operated assistant, Siri, on the iPhone of today. As the agents in the novel are based on **cloud computing**, today's devices, such as smartphones, have become redundant. They have been replaced by communication through voice iOS[11] on any device, including headphones, screens, or fixed microphones in the characters' homes.

In the future, agents will follow us everywhere, providing us with services and communication. The basic premise of cloud technology is that the sharing of computing and storage resources over the internet will create economies of scale and the ability to make complex computations without restrictions posed by individual devices or the need to invest in a massive **information technology (IT)** infrastructure.

Current **global positioning system (GPS)** technology is a good analogy for cloud computing and storage. With GPS, vehicles send their position and destination through networks to a central computer that

---

8. Jason Kincaid, "The Power of Voice: A Conversation with the Head of Google's Speech Technology," *TechCrunch*, February 13, 2011.
9. Amir Gandomi and Murtaza Haider, "Beyond the Hype: Big Data Concepts, Methods, and Analytics," *International Journal of Information Management*, 35, no. 2 (2015): 137–144.
10. Kincaid, "The Power of Voice."
11. iOS is Apple's mobile operating system.

estimates the most efficient driving route and then sends this information back to the vehicle. Similarly, when Mikhail tells *The Voice* that he wants dinner reservations, the agent connects to another agent, like today's Yelp, which stores information on restaurant reviews, prices, location, and available tables. *The Voice* then provides Mikhail with restaurant suggestions based on his stated preferences and earlier reservations. Lastly, when Mikhail approves the choice, *The Voice* connects back to the restaurant agent and makes the reservation.

Future society will be composed of a hierarchy of cloud-based agents with varying functions that collect information from devices and interact with each other. Like a head butler who leads a hierarchy of servants in pre-20th-century England estates, the personal agent will interact with many agents that fill varying functions. In the novel, these agents include *Andréas,* which monitors Mikhail's health and exercise through its collection of sensor data on heart rate, sleep, and movement.

Similarly, Mikhail has an agent, *Christina,* to handle his travels. It can connect to other agents that, in turn, are linked to timetables; data on the location of stores, restaurants, and work; and data from sensors that measure congestion and movement of people and vehicles to make real-time estimates of arrival times.

Sensor-equipped devices are an essential part of this new society, as they gather information on citizens' preferences, habits, movements, and work, providing percepts for AI-enabled agents to collect and act on.

Recent advances in **metamaterials**, materials engineered to have properties that leverage the laws of physics and energy, enable Mikhail's bedding to monitor his heart rate and body temperature, then send this information to his cloud-based health agent for analysis. Based on this analysis, his blanket's temperature will be adjusted in real time for maximum comfort.

Metamaterials are created by changing the arrangement of elements, known as meta-atoms, in conventional materials. Metamaterials gain their properties from their redesigned structures, rather than from their composition alone. This changes how the material responds to electromagnetic radiation, such as light. Metamaterials are already being used to improve performance in antennas, fiber-optic broadband, and the creation of super lenses.

A recent development is a new, low-cost metamaterial cooling film that can self-cool in broad daylight,[12] decreasing temperatures by as much of 13° C (23° F). That's enough to make a profound impact on energy consumption by cooling buildings, cars, spacecraft, and tents. This material can even be designed to self-adjust to keep a constant temperature.

Metamaterial science also allows for extreme miniaturization, thus offering the potential to create powerful antennas to fit into smartphones and power the creation of super camera lenses. Further, these materials can harvest power directly from the atmosphere.

Altogether, technologies that, until recently, we thought were purely science fiction are now possible. Researchers even are working toward creating invisibility cloaks[13] such as the one described in the Harry Potter books. This new approach to advanced material science now creates a super-low-cost platform for extremely innovative applications that were not considered feasible until now.

In the novel, the interactions we see between humans and artificially intelligent agents, and the enabling technologies that allow the collection and analysis of massive data—people's movements, social networks, preferences, health, and consumption—illustrate a likely future based on already present trends. This development is a result of commercially driven services that simplify daily life at the cost of a near-total erosion of personal data integrity and privacy.

At the heart of innovative breakthroughs, convergence will further accelerate the explosion of technology development. Just as metamaterials enable a new platform for innovation, other breakthrough concepts also will emerge.

For example, solid-state metamaterial radar has now been developed to mount in automobiles that can not only increase the resolution and identification of distant objects regardless of weather conditions, but also direct beams into buildings to gather information while disguised as 5G cellular signals. Thus, a single technology is capable of

12. Developed by PARC under the ARPA-E project. https://www.parc.com/information-sheets/metamaterials-enabled-passive-radiative-cooling-films/.
13. See https://www.parc.com/blog/design-your-own-atoms-metamaterials-at-parc/.

disrupting at least three industries: automotive, infrastructure inspection, and telecommunications.

In another example, faculty researchers at Stanford have created a new stretchable material that allows for dramatic reductions in manufacturing cost and enables a wide range of new applications for the **Internet of Things (IoT)**.

The challenge for corporate internal research and development (R&D) organizations across all industries is to become aware of these new developments and identify how emerging technologies, such as metamaterials, will likely disrupt their companies' business models.

The CEOs of Fortune 500 companies have acknowledged the importance of innovation but also recognize that they are failing to produce the breakthroughs that will drive new revenue growth. Apple's weak history of R&D since the death of its CEO and co-founder, Steve Jobs, reflects this kind of problem. Although the company is an outstanding operationally focused organization, the introduction of the iPhone X is incremental and not strategic for driving revenues forward. Apple needs to develop new categories of innovative products. Whether the company can still achieve that level of innovation remains to be seen.

TOPIC II.

PRIVACY AND THE ACCELERATION OF MOORE'S LAW

Mikhail's morning commute is a vastly different experience from those of commuters today, although it is based on already present technologies.

This future public transport system uses technologies that can track the location of individuals and then forecast commuter flows and congestion using **predictive analytics (PA)**.

This term refers to the use of historical data, statistical algorithms, and **machine learning (ML)** to estimate the probability of future outcomes.[14] Rather than just giving computer instruction about what to do, ML uses algorithms to learn and adapt through experience, with models independently changing when exposed to new data.

Currently, applications of AI and ML illustrate that it will be possible to gain knowledge about almost all aspects of society. For example, it is possible to predict voting patterns across the U.S. just using

14. This definition and some history and information on predictive analytics and machine learning appear on the website of the analytics software provider SAS.

pictures of cars. Stanford researchers found an 88 percent higher probability of a precinct voting Democratic if sedans outnumbered pickup trucks, with the opposite pattern producing an 82 percent likelihood of a precinct voting Republican.[15]

ML applied to Google Street View imaging already can predict income levels and house prices,[16] and satellite imaging produces remarkably accurate estimates of gross domestic product (GDP).[17]

Increasing amounts of available data combined with more powerful processing have led to predictive analytics transforming many business models. PA is often used to optimize marketing campaigns, detect fraud, reduce risk, and improve operations. Common uses include retailers making predictions on customer responses to marketing campaigns, cybersecurity identifying patterns in network usage to predict criminal activity, and hotels making daily predictions of the numbers of customers to maximize revenue.

We are exposed to services based on PA daily, through the movies that Netflix suggests, products presented by Amazon, and the ads displayed on Facebook. With increasing amounts of available data on our habits, consumption, and preferences, in addition to the necessary processing power and algorithms, targeted marketing will become even more accurate and pervasive by the year 2050.

In the future, our cloud-based and personal agents will manage product delivery as well. So, if Mikhail were to tell *The Voice* that he will be going shopping for new business shirts, it in turn would suggest shirts based on his earlier purchasing patterns and stated preferences. Cloud-based agents will provide information about product availability and locations, and will enable real-time purchases. Mikhail's personal agent would connect with agents linked to stores and their sales

15. Andrew Myers, "An Artificial Intelligence Algorithm Developed by Stanford Researchers Can Determine a Neighborhood's Political Leanings by Its Cars," *Stanford News Service*, November 28, 2017.
16. Edward L. Glaeser et al., "Big Data and Big Cities: The Promises and Limitations of Improved Measures of Urban Life," *Economic Inquiry*, 56, no. 1 (2018): 114–137.
17. J. Vernon Henderson, Adam Storeygard, and David N. Weil, "Measuring Economic Growth from Outer Space," *American Economic Review*, 102, no. 2 (2012): 994–1028.

systems that, in turn, are continuously determining inventory using predictive analytics.

Besides information on our consumption and preferences—traced through our social media and search history—our movements will also be monitored to an increasing degree. When these data layers are matched with our identities, goods and services will be tailored to our personal needs, although little will be left of personal privacy.

In this world, the presentation of advertisements and other information will follow people's movement across cities as well, so that billboards and displays will show individualized content, showing what is relevant for the person in the immediate vicinity, in the same way that today's personal computer screens make content visible only for the individual user. Thus, when Mikhail takes the escalator down into the subway, screens show him personalized ads, alongside information regarding his schedule for the day or public messages from the city government.

The efficiency of public services also will be supported by an increased digitalization of the environment,[18] with millions of sensors on buildings and autonomous vehicles identifying phenomena such as road-maintenance needs, traffic, pollution, utility usage, accidents, and crime, all in real time. This technology is already rapidly developing across cities. In Sacramento, California, microphones are used to recognize and track gunshots to detect crime. In San Diego, new sensor-equipped streetlights capture features such as pollution, light, and movement. In Boston, an app detects when a car hits a road bump by using smartphone sensors to identify vertical movement and sends this information to the city road maintenance department.[19] Elsewhere, sensor-equipped water pumps detect leaking pipes and can be used to identify drug usage and bacteria outbreaks through water testing.

The benefits of identifying movement have been tested and analyzed on various public transport systems,[20] typically through smart-

18. Examples of how city management is supported by digital infrastructure are provided by Glaeser et al., "Big Data and Big Cities."
19. Glaeser et al., "Big Data and Big Cities
20. A study that simulated how smartphones could be used to track public-transit usage in Chicago found that this kind of tracking could improve overall transit efficiency, with expected wait times decreased by two minutes when only 5 percent of transit users participated in the tracking system. At a 20 percent

phone applications. This allows for resources to be allotted so that capacity follows expected flows and travelers can make more informed decisions on how to travel within the city, using accurate, real-time information.[21]

Seoul, South Korea, serves as an example of smart transportation management. There the number of commuters using the subway is counted continuously. Sensors on roads monitor congestion levels, and taxis are tracked by GPS, all so that real-time information and prognosis can be provided on signboards across the city.

The London Crossrail system of 2050 takes this further, using data not only to allocate resources and the provision of information, but also to tell people when and where they can travel, a notion inconsistent with freedom of movement and democracy as we currently know it.

When paired with already available face-recognition technologies and a database linking faces with identity, it will be possible to identify everyone at a given location. This type of machine learning is already being used by Facebook, such that people in pictures are automatically identified, with an accuracy of 98 percent if they have been manually tagged in pictures several times previously.

Police cars already automatically scan license plates on nearby cars to determine if they have been stolen. From there it is a relatively small step to an application that analyzes video surveillance across a city to identify the location of any wanted person.

---

level of penetration, the average wait time went from nine to three minutes. See Arvind Thiagarajan et al., "Cooperative Transit Tracking Using Smart-Phones," *Proceedings of the 8th ACM Conference on Embedded Networked Sensor Systems* (New York: ACM, 2010).

21. Another, similar study analyzed prediction of bus arrival times based on tracking through smartphones. Knowing if a bus is going to be late based on information on other travelers, traffic, and accidents allows people to make more informed decisions on how to move within a city. The authors found that this type of prediction system provides outstanding accuracy compared to current systems. See Pengfei Zhou, Yuanqing Zheng, and Mo Li, "How Long to Wait? Predicting Bus Arrival Time with Mobile-phone-based Participatory Sensing," *Proceedings of the 10th International Conference on Mobile Systems, Applications, and Services* (Low Wood Bay, Lake District, UK: Mobisys, 2012) 379–392.

The basic premise of face-recognition technology is that each pixel in a picture is analyzed in relation to the surrounding pixels and then replaced with an arrow pointing toward the direction in which the picture is getting darker, called a gradient. The reason for replacing pixels with gradients is that although pictures vary in brightness, gradients of the same person will have the exact same representation each time.

In operation, pictures are broken down into larger squares in which the sum of the gradients that point in each major direction are counted. Each square is then replaced with the strongest arrow direction, capturing the basic structure of the image. We see one application of this operation in face detection, such as that used for automatic focus on cameras. To handle the more difficult step of recognition (rather than just detection) of faces, "face landmark estimation" is applied. Essentially, it finds 68 specific points or landmarks on a face, such as the inner edge of an eyebrow, using a machine-learning algorithm. After rescaling the picture so that the eyes and mouth are centered, it is possible to identify specific faces assuming the patterns of the face has been linked with an identity. In the case of Facebook, this is done through comparing previously tagged pictures, so if an unknown face looks similar enough to that of a tagged face, it is determined to be the same person. To simplify computing of, say, a database with millions of pictures, only a few basic measures are typically compared, such as the spacing between the eyes and the length of the nose. Machine learning is then applied to figure out which measurements are most relevant for face recognition.

An additional example of the use of ML on photos or video content is the website by Microsoft that predicts the age of people in photos (although not very accurately so far).[22] It is not far-fetched to think that eventually it will be possible to determine someone's demographic characteristics based on the clothes they are wearing.

As smartphones are network connected and equipped with cameras, microphones, and sensors, they present enormous opportunities for data collection. Although in the novel device-independent and voice-operated agents have replaced smartphones, the same functionality could be provided by any sensor-equipped device. If this data is

22. See https://how-old.net.

merged with other personal data, such as consumption data, the possibilities for tracking an individual's activities and preferences will be unprecedented, resulting in the 2050 London described.

From a commercial perspective, this will provide knowledge of the "whos" "whats," "wheres," *and* "whens" of consumption and social activity across a city, creating the opportunity to predict what someone is going to buy, where they might buy it, what their insurance and credit risks are, and even what personality type and physical illnesses[23] they might have.

Rapid technological advances now allow for the collection, storage, and processing of **Big Data.**[24] A well-known and a prevailing "truth" in the technology sector is **Moore's law**, the observation that computing power doubles approximately every 18 to 24 months. Although Moore's law originally referred only to logic chips, this development has become pervasive across technologies, with storage, algorithms, and AI developing at an astonishing pace. To illustrate this rapid development,[25] consider that only five to 10 years ago it would have been too costly to store and process vast data, while it is now possible to store information on, say, a million car rides and apply algorithms that find patterns of interest at a reasonable cost. Central to this development is cloud storage and computing, leveraging technology by providing

---

23. Current research using smartphone tracking has found that movement patterns are linked to personality. For example, extroverts move in different patterns than introverts, people who are becoming depressed will move less, and psychotic episodes are linked to erratic movement.
24. The definition of Big Data often varies, although the phrase always refers to very big datasets. A common way of describing Big Data is through three *Vs*: *Volume* (it is very large and storage would have been problematic until recently), *Velocity* (it streams and adds up at unprecedented speed, almost in real-time), *Variety* (Big Data often stems from various sources such as text, audio, and video, and is often unstructured. This often requires advanced algorithms for data processing.). Statistics platform provider SAS also adds *Variability* (data flows are often inconsistent, with periodic peaks) and *Complexity* (data stems from multiple sources, making it difficult to link and match data).
25. Donner, et al. in "Digital Cities" define this development as a quantum engine that consists of five major technological trends that drive commercial applications of Big Data: (1) rapidly increasing computing power, (2) virtually unlimited and free storage, (3) smart algorithms, (4) artificial intelligence and machine learning, and (5) advanced materials science.

economies of scale to new storage technologies[26] and to increasingly powerful computers.

Looking forward, the development of **quantum computing** will constitute an exponential leap in Moore's law and revolutionize computing possibilities. The term refers to a wholly new way of making computers, one that will provide a means to solve problems not computationally possible on traditional computers.

This type of computing is still in its infancy. It differs from traditional, transistor-based computers that store information in bits that are either zeros or ones. Put simply, in quantum computers, data is stored as zeros, ones, or any quantum superposition of those states, meaning that any state can be represented by the sum of several different states. This characteristic comes from the physical phenomenon known as **quantum entanglement**, meaning that the "quantum state" of a particle can only be described as part of a whole system. Characteristics of particles, such as their position and spin, are all correlated across the system. When two particles are entangled, neither can be changed nor measured without having an impact on the other. Basically, one particle "understands" when some type of measurement has been conducted on its entangled twin particle and the matching outcome. This works even when particles are separated by large distances—even across the universe—and without any known way for particles to share information.[27] This phenomenon, which Albert Einstein referred to as "spooky action at a distance,"[28] goes against the traditional laws of physics.

Notable is that quantum entanglement allows for putting particles in a quantum "superposition" where their properties have multiple states simultaneously, and any measurement or change of any

26. Hadoop is a platform for storing and processing enormous data. Essentially, it breaks down data into smaller pieces for processing before putting the pieces back together. This allows for parallel processing, so that the time of execution is reduced while keeping processing power, memory, and storage speed constant.
27. This is known as the EPR paradox. See Albert Einstein, Boris Yakovlevich Podolsky, and Nathan Rosen, "Can Quantum-Mechanical Description of Physical Reality Be Considered Complete?" *Physical Review*, 47, no. 10: 777–780.
28. This quote stems from the fact that quantum entanglement goes against the laws of traditional physics, causing Einstein and his coauthors to believe that quantum mechanics is incorrect or at least incomplete.

other entangled particle determines the state of the other particles. This allows a quantum computer with *n*-number of qubits to be in any superposition of 2*n* states at the same time, while normal computers can be in only one state at a time. So, a 16-bit quantum computer can store (and compute) $2^{16} = 65{,}536$ states, while a traditional computer can do only one computation. This change will constitute a true quantum leap in technology, with virtually unlimited computational possibilities. That, in turn, will allow for identification of new patterns of interest on gargantuan amounts of data, driving our knowledge about society and allowing for unprecedented levels of AI.

Quantum computing is, in fact, a technology that illustrates the rapid—and largely unnoticed—rise of China as a center for development of cutting-edge technology for both civilian and military applications. In the city of Hefei in the Anhui province, the Chinese government is overseeing a $10 billion research center for quantum applications.[29] Its two main goals for 2020 are to build a quantum computer and to realize so-called **quantum meteorology**.[30] The latter refers to new ways to measure miniscule physical effects and changes in gravity. This technology allows for accurate navigation without any external system that sends and receives signals, such as a GPS. This would allow autonomous vehicles and submarines to navigate without depending on signals that can be disturbed, blocked, or used to detect their location. Current work on quantum technology in China is focused on building a nationwide network for military communications and financial transactions, where quantum technology can provide vastly improved speed and encryption.[31]

China's advances in quantum computing can be viewed as the equivalent of producing the first atomic bomb by the United States. The technology is critical to national security in that the first nation to operationalize quantum computing is likely to become the dominant superpower of the next century. The technology in and of itself is capable of enabling intelligent self-thinking beings.

---

29. Gabriel Popkin, "China's Quantum Satellite Achieves 'Spooky Action' at Record Distance," *Science*, June 15, 2017.
30. Jeffrey Lin and P.W. Singer, "China Is Opening a New Quantum Research Supercenter," *Popular Science*, October 10, 2017.
31. See Topic Five of this companion guide for a description of quantum encryption.

The speed of processing complex data with quantum computing makes it difficult, if not impossible, for humans to comprehend. In the *First Light of Day*, speed of execution is very important, allowing Mikhail to execute his plot before anyone can respond. It becomes virtually impossible to stop what he puts in motion.

TOPIC III.

TECHNOLOGY-DRIVEN TRANSPORT

New technologies and applications have increased connectivity through free and reliable methods of communicating and spreading information, integration of financial markets, and facilitating trade of goods and services. Services once sold only locally—often requiring physical interaction—are increasingly sold on the global market, as buyers and sellers gain knowledge about each other and are provided with a means of service delivery.[32]

As with improving digital connectivity, innovation in travel will also decrease physical distances through increased speed. The supersonic commercial airplane described in the novel that Nigel flies

32. A study of nine developed countries showed that information and communications technologies (ICT) contribute between 0.2 to 0.5 percentage points a year in economic growth. See Alessandra Colecchia and Paul Schreyer, "ICT Investment and Economic Growth in the 1990s: Is the United States a Unique Case? A Comparative Study of Nine OECD Countries," *Review of Economic Dynamics*, 5, no. 2: 408-442. Similarly, the consulting firm McKinsey has estimated that the internet accounted for 21 percent of economic growth in mature economies in the early 2000s. See James Manyika and Charles Roxburgh, "The Great Transformer: The Impact of the Internet on Economic Growth and Prosperity," McKinsey Global Institute, October 2011.

between London and Chennai is currently being developed by Boom, and is being scheduled for a first test flight next year. Capable of speeds slightly faster than a Concorde at Mach 2.2, a flight on **Boom**, as it is called, between London and New York will take three hours and 15 minutes, rather than the seven hours of traditional commercial airplanes. This at prices equivalent to a business-class ticket on a traditional airplane, and with a sonic boom—the sound blast that made the supersonic flight unsuitable for flight over land—30 times quieter than that of the Concorde.

This type of flight is only one end of the spectrum of future air travel, likely to be reserved for the affluent and time-pressed. In addition, it's not exactly environmentally friendly, though fuel-burn per seat and mile is also equivalent to today's business class.

A current project that is likely to illustrate the future for much of the commercial air industry is the electric **Airbus E-Fan** program for electric and hybrid airplanes. In 2014, a two-seater plane equipped with lithium-ion batteries powering two electric motors had its initial test flight, and a longer-range hybrid version debuted in 2016. Current work toward a commercial regional airplane is underway, with 200 Airbus and Siemens employees dedicated to working on demonstrating the technical feasibility of hybrid airplanes by the year 2020.

The modern history of the aviation industry is a good illustration of the way sensors and the ability to gather and analyze data with predictive analytics is changing business models, with an increasing emphasis on service delivery rather than physical construction. For example, airplane engine– maker Rolls Royce now sells the service of providing, say, 10,000 hours of engine operations rather than selling the actual physical engine.

This is made possible by sensors that monitor engine performance and then send data back to the manufacturers, something that is now prevalent across most transportation industries.[33] Network-connected cars and airplanes continually exchange information with their manufacturers. For example, Tesla continually performs over-the-air software updates of its cars.

---

33. This type of massive data and the ability to optimize performance have led to a new challenge for airplane engine manufacturers: data on failures is now so rare that their algorithms for predicting failure are not optimized.

Just as Rolls Royce has transformed the way it sells airplane engines, automakers will move from being in the business of selling cars to providing the service of transportation.[34] Car ownership will become increasingly less important for people, with transportation being sold either on the aggregate vehicle level through car-sharing services such as Zipcar or on the level of individual trips such as with UberPool or Lyft Line.

A focus toward price per mile and marginal cost of use, rather than the up-front, fixed cost of buying a car, is going to be amplified by electric vehicles that offer significantly longer mileages, as the high cost of batteries can be amortized over many trips. Added to this are sensor technology and AI that make **autonomous vehicles (AV)** a reality. That revolution will decrease costs for car-sharing services by about 70 percent per mile.

It is likely that car-sharing services will become the major buyers of cars, which will change business models and depress margins for automakers. The shift towards per-mile pricing will increase car usage and lead to more rapid replacement cycles, meaning that new technology will be implemented at a faster pace. Cars also will be designed with an increased focus on passengers, as shown in the novel, when Mikhail communicates with his personal agent, *Christina*, through the car's audio and video system.

AVs currently tested by Google are equipped with several sensors: a rotating sensor that creates a constantly updated three-dimensional (3-D) map of the environment, a camera that detects traffic lights and moving objects, a sensor that measures sideways movement and the vehicle's position on the map, and four radars that measure distance to identify any kind of obstacle. AVs navigate using radars like those on airplanes, emitting radio-frequency waves to determine how far away an object is.

Metamaterial[35] and AI development also are rapidly improving AV technology. Engineered materials now enable highly directed radio-frequency beams that produce a true 3-D image of the surrounding environment, determining the location and speed of all

34. Bill Ford, "Bill Ford on the Future of Transportation: We Can't Simply Sell More Cars," *Wall Street Journal*, July 7, 2014.
35. See Topic One of this companion guide, on metamaterials.

surrounding objects, even in bad weather and cluttered environments. This gives the AI better percepts so that the vehicle can "see" around corners, apply predictive analytics of traffic patterns, identify dangerous situations, and prevent accidents before they occur.[36]

36. The company Metawave has developed radar technology with metamaterials (known as the WARLORD radar), as well as AI solutions for autonomous vehicles. See https://www.metawave.co.

TOPIC IV.

SELF-CONSCIOUS ARTIFICIAL INTELLIGENCE

The meeting between Nigel and Tharra, and their discussions on the meaning of self-consciousness with the Indian guru Chandrashekhar Sekhar (CS), illustrate that artificial intelligence is unlike any other new technology.

AI enables new analysis of data by finding patterns in enormous data caches beyond any human's limited cognitive ability. AI faces no such limitations. While humans have always created tools that enable us to create great things, AI is giving us the capabilities to potentially create something with its own will and cognitive abilities far beyond our own.

As many have stated, the impact of this revolution is so enormous that it is impossible to predict what can be achieved when artificial intelligence surpasses human intelligence.[37] For example, AI is already able to identify skin cancer through smartphone pictures, with higher

37. Stephen Hawking et al., "Transcendence Looks at the Implications of Artificial Intelligence—But Are We Taking AI Seriously Enough?" *Independent*, May 1, 2014.

accuracy than 21 experienced physicians.[38] Increased knowledge of our DNA, paired with AI, predictive analytics, and robotics, can revolutionize health care through prediction and treatment of disease, all performed by robots at a low cost.

The potential downside of AI does, however, require attention, as illustrated by weapons manufacturers who are already planning for autonomous weapons that select and attack targets. Similarly, AI will have a profound impact on labor markets and wealth inequality, with many jobs becoming redundant.

Thus, when CS cautions about the unknown consequences of self-conscious AI, as developed by the company *M*, he is, in fact, in agreement with many of today's technology leaders who raise concerns about the potential dangers of AI.[39]

What is especially worrisome are the unintended consequences that might follow when machines become "super-intelligent." On this, a group of world-leading scientists state: "One can imagine such technology outsmarting financial markets, out-inventing human researchers, out-manipulating human leaders, and developing weapons we cannot even understand. Whereas the short-term impact of AI depends on who controls it, the long-term impact depends on whether it can be controlled at all."[40]

In addition to these risks, providers of cloud-based agents—in this case the fictional company that delivers His Master's Voice—will have unprecedented knowledge about people and will control what information is presented to them. We already see this trend reflected in today's

---

38. Andre Esteva et al, "Dermatologist-Level Classification of Skin Cancer with Deep Neural Networks," *Nature* 542, no. 7639: 115–118.
39. In January 2015, dozens of AI experts signed an open letter on the dangers of AI. The letter has received more than 8,000 signatures to date. See Stuart Russell, Daniel Dewey, and Max Tegmark, Association for the Advancement of Artificial Intelligence, "Research Priorities for Robust and Beneficial Artificial Intelligence: An Open Letter," *AI Magazine*, Winter 2015: 105–114. See also: Hawking et al., "Transcendence Looks at the Implications of Artificial Intelligence"; Holley, "Bill Gates on the Dangers of Artificial Intelligence"; Peter Holley, "Stephen Hawking Just Got an Artificial Intelligence Upgrade but Still Thinks It Could Bring an End to Mankind," *Washington Post*, December 2, 2014.
40. Hawking et al., "Transcendence Looks at the Implications of Artificial Intelligence)

business models, which are based on identifying people's needs and preferences and knowing what products to sell to which customers.

In fact, information is rapidly becoming a new asset class, with personal data as the "oil of the 21st century."[41] Take Google, for example. In addition to knowing what users search for, Google also has the ability to decide what to present to users. Using similar capabilities, foreign governments, through Facebook, may have influenced the U.S. presidential election of 2016.

When personal agents begin to handle our travel, work, consumption, communication, and searches for information, collecting and analyzing more data and thereby creating a fuller picture of our personal characteristics, this type of influence will be amplified. It is not unthinkable that commercial interests will influence these services. For example, they might provide us with a route to work that passes by sponsored retail locations, or the agents might frame sentences to influence our opinions in some way.[42] So besides leading to the end of privacy, new technology might, subconsciously, influence our decision making.

Consumer demand for new technology has created a **democratization of technology** phenomenon. Large enterprises are now influencing government policy based on their ability to use consumers to circumvent government policies. Apple's refusal to deliver content to the National Security Agency (NSA) and other government organizations exemplifies the power of this new phenomenon.

Finally, we should say a word about the role of government and cities in this new world. The largest, richest, and most heavily targeted markets in the world are in cities. Commercial companies, including Apple, Facebook, Uber, and others, are spending enormous R&D budgets developing applications for urban markets. City, state, and federal government spending on the digitization of cities pales in comparison.

---

41. World Economic Forum, "Personal Data: The Emergence of a New Asset Class: An Initiative of the World Economic Forum," World Economic Forum, January 2011.
42. See, for example, the "framing effect," a term from behavioral economics, which shows that choices are impacted by either highlighting the positive or negative in the same decision. Daniel Kahneman and Amos Tversky, "Prospect Theory: An Analysis of Decision Under Risk," *Econometrica*, 47, no. 2 (1979): 263–291.

It is the commercial companies that will decide how citizens in cities will function and what applications they will use. Most of our city governments are legacy-based systems that were designed for the immediate demands of the postwar era before technology exploded to the forefront. As such, their policies, expertise, and aptitude are woefully unprepared to deal with the new world realty. Perhaps the best example of this is Silicon Valley, the heart of global innovation and technology risk taking.

Here in Silicon Valley, we have terrible traffic congestion, a failing infrastructure, city governments that do not coordinate to solve regional problems, and a near-complete lack of expertise in city planning for future needs. In fact, along with incompetence in city planning comes a culture of arrogance and alarming inequality. Silicon Valley has, among other things, one of the highest poverty rates in the country, the most expensive housing, some of the worst infrastructure costs, average public education, and high rates of homelessness. No matter how advanced our technological knowledge is, social and government expertise is way behind the curve and not likely to catch up.

TOPIC V.

ENCRYPTION AND CROSSING THE DATA LAYER

London of 2050 is not just a physical city of roads and buildings; it is also a collection of 3-D commercial and city data layers, such that every aspect of human existence within cities can be analyzed, understood, and predicted for future advantage.

Much of the technology represented in this depiction of London 2050 already has been described in this companion guide, most notably sensors. Today, sensor data is often collected—although not yet centralized—for entire cities or crossed with other data, a development that will create true value, leading to a real-time digital representation of almost every aspect of a city.

Current research on merging datasets will lead to the ability to visualize an entire city and analyze "What if?" scenarios using predictive analytics.[43] For example, consulting company WSP has developed

43. This is part of the "Digital Cities" project at Stanford University, taking place at the Global Projects Center at the School of Engineering. See Donner, Eriksson, and Steep, "Digital Cities: Real Estate Development Driven by Big Data," for a discussion of urban applications of Big Data.

a modeling platform of the entire city of Seattle. It is used to simulate the impact of an earthquake based on data about durability for all buildings and infrastructure.

The power of this platform comes from its visualization of outcomes. Using highly credible data, it can show what an earthquake will do to the city. This same technology can be used to drop new buildings into the model to see the impact on traffic, power, consumer locations, and building return on investment. Any data type may be used in the model, such as commercial transactions or movement.

A viewer of the model can "fly" over the city of Seattle or dive underneath the city to see the utilities, power lines, and water-delivery systems. Eventually, the technology will make it possible to look at city economics in a new and dramatic way, seeing how money moves through the city, where it stops, where it grows, and where it exits. Commercial companies will be able to leverage this experience to make consumer-facing marketing decisions and to develop commercial markets within urban locations.

As data crosses the layers of the city, it will be combined to create a new, in-depth profile of who we are; where we work; and where we shop, recreate, and entertain. The financial power of the model will test our ability to maintain the legacy concept of individual privacy.

Even today, we see rapid expansion underway in the availability of various kinds of information, such as digitized public records, so that data on matters such as crime, taxes, and education can be immediately applied to monitor the impact of policy. For example, law enforcement could use real-time updates on crime within very small geographies.[44]

Commercial applications and social media also provide vast opportunities when crossed with other data.[45] A dataset combining demographic information, movement, and preferences for every individual in a city can tell us what kind of services, products, or housing people are most likely to want or buy, and where these individuals live and spend their time. The real-estate listing service Zillow already provides information on housing preferences, and the restaurant-

44. Anthony A. Braga and Brenda J. Bond, "Policing Crime and Disorder Hot Spots: A Randomized Controlled Trial," *Criminology*, 46, no. 3: 577–607.

45. See Glaeser et al., "Big Data and Big Cities," for an overview on how Big Data can be used in cities.

booking-and-review platform Yelp tells us the preferences of demographic groups and the location of various types of establishments. In fact, restaurant reviews are currently used to allocate scarce resources for health-and-safety inspections in New York, so that restaurants with bad reviews are checked more frequently.[46] Similarly, the location and timing of Google searches can tell us where outbreaks of influenza are taking place so that public-health resources can be more efficiently directed. Now imagine these and other services becoming more capable by orders of magnitude when powered by quantum computing.

Sensors on buildings and on handheld devices can tell us the number of people at a certain location, in what direction they are walking, and what kind of activities are taking place (such as waiting in line, eating, drinking, or walking). In the future, data on movement from smartphones could tell us that people who live at location A often work in location B and then eat lunch at location C, providing valuable insight for new real-estate and infrastructure development, marketing efforts, and public transport.

This type of data gathering raises issues concerning personal privacy and data integrity, and some sort of data **encryption** is vital. The basic premise of encryption is to scramble data so it is illegible for unintended parties. This kind of data protection goes way back. Even Julius Caesar camouflaged important messages with encryption. Today, the key between the actual information and the encrypted data, **the cipher**, is created through complex algorithms, with key size and strength as the main differences among various technologies. Early data encryption often followed the common **Data Encryption Standard**, which was adopted by the United States in 1976 and subsequently spread across the world. A variety of faster and more secure subsequent standards have emerged since the late 1980s.

Modern encryption takes two primary forms: **symmetric key algorithms** and **asymmetric key algorithms**. The former is comparable to multiple users having identical keys to the same lock, while the latter means each user has a unique key. With the latter, when person A sends information to person B, it is sent with a lock that only B can open—and vice versa for the reply. Basically, two sets of keys are used,

46. See Glaeser et al., "Big Data and Big Cities."

one public and one private, so that anyone can encrypt information with the receiver's public key. The receiver then decrypts the information with his or her private key. This type of encryption prevents third parties from gaining access to the key and prevents security breaches from spreading (i.e., if A's key becomes known, it does not provide access to B's data).

Although generally considered to be secure, asymmetric encryption technology is dependent on the amount of computing power someone trying to break the encryption has at his or her disposal. Current supercomputers can now crack standard encryptions, and rapidly increasing computing power will only further accelerate the need for better—supercomputer-proof—encryption.

Just as quantum computing will revolutionize our ability to process data, **quantum mechanics**[47] will also transform encryption.[48] The basic (and still somewhat complex) premise of **quantum encryption** is that the key is incorporated into a photon—a light particle—that is correlated with a second photon in quantum entanglement.[49] This means that any attempt to measure or observe one photon impacts the other. The photon in which the key is embedded is sent to a receiver through fiber cables,[50] making it unlikely that someone could break into the data when it is between the sender and receiver on the network. Since the pair of photons are entangled, any such interruption would alert the sender.

One obstacle is that it is difficult to create databases that are both encrypted and usable. That said, researchers recently have found a fast and efficient way to make computations without ever decoding the data.[51] In the novel, for example, Mikhail addresses the privacy con-

47. This is also known as quantum physics or quantum theory and refers to the description of the smallest levels of energy on the levels of atoms and subatomic particles. This contrasts from classical physics, the theories of which describe nature at its normal scale.
48. See Clay Dillow, "Unbreakable Encryption Comes to the US," *Fortune*, October 14, 2013.
49. Quantum entanglement is described in greater detail in Topic 3 of this companion guide.
50. This makes this technology heavily dependent on the quality of fiber networks.
51. This is known as "fully homomorphic encryption." See Andy Greenberg, "An MIT Magic Trick: Computing Encrypted Databases Without Ever Decrypting Them," *Forbes*, December 19, 2011.

cern raised by Tharra regarding data that tracks schedules, work habits, home behaviors, and transactions of all London residents. It is this new technology that allows for query-into-encrypted-data computations on encoded strings of numbers that yield decoded results identical to those that would be produced if the data hadn't been encrypted at all.

It is, however, important to note that encryption does not imply privacy. The main drive for encryption is to protect the value of the information, rather than the privacy of individuals. In the scenario depicted in the *First Light of Day*, depersonalized encrypted information is used to target potential customers who subsequently voluntarily opt to turn over their information to the service provider, similar to the terms and conditions of most commercial and social media applications of today. Then, as now, most people voluntarily provide various companies with information such as their travel, consumption, and income every time they use a service.

TOPIC VI.

CYBERSECURITY AND WARFARE

New vulnerabilities emerge because of our increasingly digitalized society, part of the so-called **Internet of Things (IoT)**.

IoT refers to network-connected devices such as smartphones, vehicles, sensors, home appliances, and even infrastructure such as water pumps, that send and receive information. This development is exponential, with the number of internet-connected devices predicted to reach 200 billion by the year 2020,[52] compared to 15 billion in 2015 and just two billion in 2006. Growing in parallel, cybercrimes are estimated to have cost the global economy more than $450 billion in 2016, with two billion personal records ending up in the hands of criminals.[53]

This is not only a commercial problem, for as society is becoming increasingly reliant on technology, the ability to damage or destroy digital infrastructure will become an increasingly important part of warfare. Records and services central to the work of government, military, and financial entities are also increasingly digitalized. The advent of cloud computing and Big Data increases incentive for attacks. We

52. This is predicted by the microprocessor manufacturer Intel.
53. Luke Graham, "Cybercrime Costs the Global Economy $450 Billion: CEO," *CNBC*, February 7, 2017.

see the associated risks in the novel—Russia's spying on both His Master's Voice and the personal information of millions of people, provided through its interaction with personal agents.

The basic premise of cyberattacks is to exploit computing vulnerabilities related to what is known as **information assurance (IA)**, namely the process of storing and processing data and ensuring that the correct information is provided to the right person at the right time. The field of cybersecurity is concerned with three principles[54]: first, that a flaw in the system is subject to attack; second, that the attacker has insight about this flaw; and third, that the attacker has the capabilities to exploit that flaw.

Attacks on computer networks are typically initiated by the installation of malicious software or **malware,**[55] which can provide hackers with access to the user's information, such as banking, passwords, and files. Alternatively, malware can destroy data, cause system failures, and drain computing resources.

The initial installation of malware typically requires some sort of human action, such as so-called phishing, where attackers attempt to get sensitive information such as passwords, by sending emails asking for information or linking to counterfeit websites. As malware is typically designed to spread across networks, increased connectivity leads to greater risks associated with malware infections.

Cyberthreats are posed by individuals, groups, and governments aiming to shut down a nation's networks and computers, to disrupt delivery of essential services, or to steal information or money. Criminals often steal identities and personal information to use in

---

54. Don Snyder et al., "Improving the Cybersecurity of U.S. Air Force Military Systems Throughout Their Life Cycles" (Santa Monica, CA: RAND Corporation, 2015).
55. Malware are typically categorized as either Trojan horses, computer viruses, or worms. Trojan horses refer to programs that claim to be something that they are not, so that one installs the program (often by downloading an attachment or clicking on a banner) unaware of its true and malicious nature. Trojan horses are often used to gain access to a computer system, causing loss or theft of data. Computer viruses are programs that change the behavior (i.e., code) of other "infected" programs and replicate themselves through other programs. Worms are like viruses with the main difference being that worms do not need an infected "host file" to spread, unlike a virus. Computer threats are often blended (i.e., a mix of these three malicious program categories).

illegal activities, such as to make online purchases, blackmail victims over sensitive information found on targeted computers, or commit so-called **cyber extortion**, shutting down a system and demanding money in return for restored access.

An example of this is the malware **WannaCrypt**, which exploited a vulnerability in Microsoft software on more than 75,000 computers in 99 countries. It allowed for the encryption of content on victims' computers. The attackers subsequently required payment of $300 worth of bitcoins to decrypt the files so that the user could regain access to the information.[56]

Another major cybercrime was committed by a criminal organization known as **Carbanak**, based in Russia, China, and Europe. From 2013 to 2015, this organization stole an estimated $1 billion from numerous banks through the installation of malware that tracked every move on bank computers that handled bookkeeping and wire transfers.[57]

It all started with bank employees installing the malware through emails that looked like normal office correspondence. The malware went undetected, while intercepting the screens of bank employees and sending back images and videos that gave the criminals insight into bank routines and systems. The criminals even managed to maintain continuous access to the bank system through remote access tools similar to those used by IT support functions.

Being able to make transactions—and make them look normal—enabled the criminals to transfer funds between accounts and to order cash to be dispensed from ATMs. This strategy is commonly used to enable penetration of the first line of network defense, **firewalls**.[58]

---

56. Robert Hutton, Jeremy Kahn, and Jordan Robertson, "Extortionists Mount Global Hacking Attack Seeking Ransom," *Bloomberg Politics*, May 13, 2017.
57. See David E Sanger and Nicole Perlroth, "Bank Hackers Steal Millions via Malware," *New York Times*, February 14, 2015, and James Cook, "This Is Exactly How a Gang of Incredibly Patient Hackers Stole up to $1 Billion from Banks Around the World," *Business Insider*, February 16, 2015.
58. A firewall can be hardware, software, or both. Various types of firewalls exist, such as proxy firewalls, which serve as gateways between networks (offering protection by preventing connections from outside the network); stateful inspection firewalls, which monitor all traffic through a connection while it is open, with filtering decisions made by set rules and administrators; and unified threat management (UTM) firewalls, which combine firewall capabilities with

These have the primary function of monitoring incoming and outgoing traffic, and deciding whether that traffic should be allowed. When hackers gain insight into organizational routines, firewalls become a less efficient means of protection. Because Carbanak's transactions looked like normal transactions and followed regular procedure, they could get past the firewall.

Hackers can also gain access to data by bypassing normal authentication through what is often referred to as a **backdoor**, typically included in systems either intentionally, by mistake, or by installation through attack. In fact, in 2013 it was revealed that a backdoor was installed by the U.S. Department of Defense's National Security Agency into the National Institute of Standards and Technology (NIST) encryption standard[59] supporting global surveillance of Google and Yahoo accounts, phone records, email, and instant-messaging content.

Data security is a perpetual arms race between developers and hackers, illustrated by the recent revelation that Intel processors—used in most of the world's computers, tablets, and smartphones—have two major flaws (known as **Meltdown** and **Spectre**), which can potentially provide hackers with access to their memory. This is especially problematic for cloud storage providers, with hackers being able to access information such as customers' passwords. Current information suggests these flaws can be addressed with a so-called **patch**, which is a piece of software that updates or remedies a problem with a computer program or supporting data. However, such patches can decrease computer performance by as much as 30 percent.[60] Of the two flaws, Spectre is the more problematic. It might not be possible to address this flaw without a full architectural redesign of the processors themselves—meaning that the computer might need to be replaced.

Besides attacks by criminals with financial motives, nations are increasingly using cyberwarfare in conflicts. A prominent example is the malware **Stuxnet**, which is thought to have destroyed upwards of

---

antivirus software to prevent, detect, and remove malicious software. Newer-generation firewalls now focus on identifying the greatest risks within a network and quickly identifying and reacting to suspicious activity within the network.

59. Kim Zetter, "How a Crypto 'Backdoor' Pitted the Tech World against the NSA," *Wired*, September 24, 2013.
60. Cade Metz and Nicole Perlroth, "Researchers Discover Two Major Flaws in the World's Computers," *New York Times*, January 3, 2018.

one-fifth of Iran's nuclear centrifuges before being discovered in 2010. It is the first known use of cyberweapons and is likely to have been installed through USB flash drives. In operation, it changed the centrifuge controls so that the fast-spinning centrifuges tore themselves apart.

Although not confirmed, Stuxnet is widely believed to be part of a U.S.-Israeli operation known as "Operation Olympic Games," aimed at sabotaging the Iranian nuclear program, an operation that is thought also to have included the so-called Flame computer worm that infected computers across the Middle East.[61] Masked as a Microsoft software update, the code quickly spread across highly secure networks and went undetected for years, all the while controlling computer functions such as activating microphones and cameras, logging keyboard strokes, taking screenshots, and transferring stolen information to distant hackers.

Warfare of the future will be increasingly fought online, with large-scale cyberattacks having the potential to obstruct access to the services of banks, governments, and communication technology, all of which would wreak havoc in every corner of society. It is theoretically feasible for hackers to gain control of airplanes, flight-control systems, and power grids.[62] Experts are raising the alarm that current security levels and preparation for cyberattacks are far from sufficient.

A scenario such as a large-scale attack on the U.S.—resulting in inaccessible online banking sites, disabled ATMs, and nonfunctional internal accounting systems—would paralyze the nation's economy. And the perpetrators would be all but untraceable.[63]

Complicating matters, in cyberwarfare, the likelihood of unexpected consequences and collateral damage are much higher compared to that associated with traditional weapons, as malware is typically designed to replicate and spread across networks. A cyberattack on

---

61. Ellen Nakashima, Greg Miller, and Julie Tate, "U.S., Israel Developed Flame Computer Virus to Slow Iranian Nuclear Efforts, Officials Say," *Washington Post*, June 19, 2012.
62. Kevin Loria, "Cybercrime Poses a Potential Existential Threat to Our Society, and We're Completely Unprepared," *Business Insider*, May 22, 2015.
63. Christopher S. Chivvis and Cynthia Dion-Schwarz, "Why It's So Hard to Stop a Cyberattack—and Even Harder to Fight Back," *Rand Corporation Blog*, March 30, 2017.

the power supply at a specific location might very well spread beyond its intended target and shut down the electrical supply for an entire region.[64]

In fact, Stuxnet was intended for a specific Iranian network, yet it was discovered only after it spread beyond its intended target. It did not conduct further attacks because a self-destruct date was included in the code. Had the hackers been less careful, the attack would have been far more widespread. Of course, the potential for accidental conflict only increases when it is difficult to understand the intent of attacks and the consequences of retaliation.[65]

These technologies, strategies, and cybersecurity concerns come into play in the novel, which depicts a future where increasing amounts of personal data, connectivity, and cloud storage have intensified the arms race between developers and hackers.

---

64. This example came from Chivvis and Dion-Schwarz, "Why It's So Hard to Stop a Cyberattack."
65. Chivvis and Dion-Schwarz, "Why It's So Hard to Stop a Cyberattack."

TOPIC VII.

HUMAN-MACHINE INTERACTION

Current technologies and the advent of personal agents will have a profound impact on how humans communicate with one another and with machines.

Much of the future depicted in the *First Light of Day* illustrates how the line between real, human interaction and comparable interactions with technology will be increasingly blurred. For example, texting between humans will be handled by AI, and daily interactions between people will be managed by agents whose actions are almost indistinguishable from those of humans.

Day-to-day activities that used to require interaction between humans will now be completed by machine, spanning everything from making purchases online to ordering fast food through a touchscreen to ordering drinks in a bar after work. This development will continue as advances in AI lead to algorithms that not only, like today, recommend products on Amazon and video content on Netflix (acting on our behalf), but also offer more advanced services such as bookkeeping, taxes, and basic legal work.

Efficiently giving instructions to machines and in turn receiving useful information from them is part of a field of research known as

human-machine interaction (HMI).[66] Broadly, HMI encompasses studies on the design, evaluation, and implementation of systems for human use. Current initiatives in this area include improving the interaction between humans and technology. In the case of a computer system, for example, much work centers on the usability of systems and the computer's interface.

By nature, this is an interdisciplinary field, relating technology to questions of culture, values, and social norms that can and should impact computer systems. Part of this is so-called **augmented reality (AR),** which means that layers of information are added to our perceived reality, such as a system that adds information about the people we are talking to on a screen or through AR eyeglasses, such as Microsoft HoloLens or Google Glass. This is slightly different from the better-known **virtual reality (VR)**, which refers to a fully virtual experience rather than one that is only partially augmented.

Research on such eyewear is part of a subfield of AR that specifically refers to changing perceived reality through wearable computers or handheld devices, a subfield known as **computer-mediated reality (CMR).** Such AR-enabling devices have much in common with the AI-based agents depicted in the *First Light of Day.* Microsoft's virtual agent, Cortana, for example, is one of several applications for the HoloLens.

Various applications also use an AR display's ability to display holograms. The "Sidekick" project, a collaboration with the National Aeronautics and Space Administration (NASA), is one such application. This initiative will enable a HoloLens to transmit the identical perceived reality the astronaut sees, in 3-D, to the ground crew, who in turn can give the astronaut instruction and support by adding illustrations to the image. This system is currently being deployed on the International Space Station, replacing printed instruction and voice-based communications. It is increasing the efficiency of complex tasks associated with space exploration and potentially decreasing the need for training.

66. Also known as human-computer interaction (HCI), man-machine interaction (MMI), or computer-human interaction (CHI).

A commercial application of this technology is HoloTour, which allows for realistic 3-D virtual tourism that creates the experience of moving through any place without actually traveling there. Google Earth is another well-known, primitive example of this type of VR, offering a similar experience in which one can walk through any city at the street level.

In the novel, most of Mikhail's interactions are with AI agents: chatting with his personal agent about his schedule, work, and shopping; gaming his social life through texts he's not actually writing himself; and experiencing a therapy session in which AI and ML enable a realistic interaction with his dead childhood friend, Sasha.

This is not science fiction. VR already is being used by psychologists as a cost-efficient way to study human behavior, including such phenomena as how humans react to "being in someone else's shoes," how digital influences alter perception and emotions, and how technology has changed social interaction.[67] Other uses for VR include training for doctors, flight simulation for pilots, and the training of astronauts while still on earth. The latter, an application of VR NASA has used for more than 20 years, simulates zero-gravity environments.

Understanding how interaction occurs in a virtual environment has widespread importance for applications built on this type of technology. Perceived trustworthiness becomes hugely important, implying that AI-based agents should be made more attractive and life-like to make humans trust their instructions or, in the case of commercial transactions, buy recommended products or services. Adding to the overall complexity is that exposure to virtual or augmented reality is likely to impact our psyches in ways not fully understood, as little research on this impact exists. It is likely that long-term exposure—where the line between reality and "synthetic" reality is blurred—will ultimately result in physical disorders and impaired decision making when real-life choices are based on virtual experiences.

The increasing number of tasks being conducted by machines raises the question of how much we should trust technology. Some research on human-machine team performance is already aimed at calibrating

---

67. The field of study on how social interaction occurs in virtual environments is known as transformed social interaction.

the appropriate level of trust that should be given to a machine,[68] given that AI also makes mistakes, especially in settings characterized by uncertainty and doubt.

How should the intent and honesty of an AI-based agent or robot be determined? This question will become increasingly important given that many, if not most, financial, medical, and legal issues will be handled online. In fact, most of us feel far more comfortable giving up information in an online questionnaire than we do in an interaction with an unfamiliar human—something that often is far from rational, given the commercial use of personal data.

---

68. This research is a collaboration between the Massachusetts Institute of Technology and Singapore University of Technology and Design. Jesse DeLaughter, "Building Better Trust Between Humans and Machines," *MIT News Online*, June 21, 2016.

TOPIC VIII.

CRYPTOCURRENCIES

The way we conduct transactions in the future will be very different from how we do so today. These changes will have an impact on when and where we spend money—for example, making a purchase through a conversation with your personal assistant in an autonomous vehicle on your way to work.

These shifts will include the means of payment. Government-backed currency transferred through cash, credit, or debit card is the typical transaction of today. Traditional currencies—such as the U.S. dollar, the euro, the Chinese renminbi, and the Japanese yen—are "free-floating fiat currencies," meaning that exchange rates are determined on the open market, and their underlying value comes from being backed by a central bank promising payment and from a legal system to resolve conflict. Thus, a U.S. dollar is basically an IOU backed by the U.S. Federal Reserve.

This type of currency, backed by a central bank with a monopoly on money supply, has been the prevailing system, with alternative currencies (not recognized by the legal system as mediums of exchange, or so-called legal tender) having had limited use. But the advent of **bitcoin** in 2009 has changed everything.

Bitcoin is one of the first and perhaps the most famous of **cryptocurrencies**, which offer a digital asset and medium of exchange, and an alternative to traditional currencies. At the heart of cryptocurrencies is cryptography—the encryption that secures the supply of new units of currency and the verification of transactions, thus providing heightened trust to users.

The backbone of cryptocurrencies is **blockchain**, a public ledger of all bitcoin transactions. The system is decentralized and supported by "nodes," which are computers connected to the blockchain network that administer the task of verifying transactions added to the blockchain.

Each user is identified with a unique number, known as a blockchain address. It is between these addresses that transactions are executed. Once a transaction has occurred, records are impossible to change, as each transaction is linked to the previous transaction. (That is the "chain" part of blockchain. "Block" refers to groups of transactions that are continuously added.)

Validation[69] of blocks of transactions is handled by administrators who, in return for conducting this task, compete to win bitcoins by solving mathematical problems. This practice, known as mining, is how new units of cryptocurrency are created. In the case of bitcoin, the maximum number of units is set at 21 million, with a supply of about 16.6 million coins available by the end of October 2017. Thus, where supply of traditional currency is at the discretion of central banks and subject to change based on economic necessity, a cryptocurrency has a predetermined upper limit to supply.

Anonymity and lack of centralized control do, however, pose issues, most notably a concern that cryptocurrencies enable illegal activities, including tax evasion, money laundering, extortion, and bypassing trade restrictions. In addition, there is the risk of theft. North Korea often attempts cryptocurrency theft through cyberattacks on cryptocurrency exchanges.[70] In terms of anonymity, so-called altcoins are highly preferred by criminals, as any identifying information is

---

69. This design is intended to keep bitcoin safe by preventing "double-spending," i.e., someone spending the same bitcoin twice by illicitly making copies of it.
70. Yuji Nakamura and Sam Kim, "North Korea Is Dodging Sanctions with a Secret Bitcoin Stash," *Bloomberg Businessweek*, September 11, 2017.

removed from the blockchain of transactions.[71] In relation to the novel, cryptocurrency enables illegal and clandestine activities.

Blockchain allows digital information to be distributed but not altered or copied. The technology therefore has the potential to offer a transparent and predictable way to store not only cryptocurrency but also any type of sensitive data. If contracts are embedded in code and protected from being changed or deleted, and signatures are identified and validated via blockchain, the need for intermediaries, such as lawyers and brokers, will diminish.[72] This does, however, raises a serious concern. If the technology proves to be less perfect than it is currently believed to be, dependence on it might have catastrophic consequences.

---

71. Jason Bloomberg, "Using Bitcoin or Other Cryptocurrency to Commit Crimes? Law Enforcement Is onto You," *Forbes*, December 28, 2017.
72. Marco Iansiti and Karim R. Lakhani, "The Truth About Blockchain," *Harvard Business Review*, January–February 2017.

TOPIC IX.

CLIMATE CHANGE

Climate change and global warming[73] will have a profound impact on society, likely resulting in large-scale migration, land loss, and wealth redistribution as land, infrastructure, and buildings become uninhabitable.

Although the climate has changed throughout earth's history, with the last ice age 7,000 years behind us, most of these historic changes have occurred due to shifts in the earth's orbit, varying the amount of sun our planet has received. Recent changes, however, have been extremely rapid. The temperature has increased 2° F (1.1° C) during the 20th century, with most of that change occurring during the past 35 years. In fact, 16 of the 17 warmest years on record have occurred since 2001. During 2016, eight respective months experienced their highest temperatures on record.[74] This increase in temperature is due to the heat-trapping nature of carbon dioxide and other gases, the

---

73. NASA's website provides a thorough overview of the evidence of climate change, its causes, its effects, and ways to tackle it: https://climate.nasa.gov/evidence/.
74. See https://climate.nasa.gov/evidence/.

greenhouse effect, which we have known about since the mid-19th century.[75]

There is no scientific doubt that greenhouse gases are the root cause of climate change,[76] a consequence of human behavior, and that future developments will be impacted by human behavior.[77] Even the most conservative estimates of future climate change will require massive adaptation to new circumstances, with more extreme weather, changed sea lanes, and mass extinction of species.

Already, we are experiencing rising temperatures on land and in oceans. Ice sheets in Antarctica and Greenland are melting, as are glaciers on mountaintops elsewhere across the planet. Seasonal snow covers are decreasing. And sea levels rose by eight inches during the20th century, with the rate of increase doubling during the past two decades.[78]

A major theme of the novel is that technology is developing at an exponential pace and that humans, who tend to think linearly and in our comfort zones, can't keep up. Climate change is another example of the human inability to handle rapid change, illustrated by our lack of preparation for rising sea levels and more extreme weather.

A good illustration—although a far from unique case—is the San Francisco Bay Area, home of Google, Facebook, and Apple, and research institutions such as Stanford and the University of California, Berkley. There investors seem to have ignored climate change—with 27 large construction projects, totaling $21 billion in value, proposed in areas vulnerable to flooding in the coming decades.[79] Between 2010

---

75. Physicist John Tyndall recognized the so-called greenhouse effect and its potential impact on climate in the 1860s. In 1896, Swedish scientist Svante Arrhenius predicted that higher atmospheric levels of carbon dioxide could increase surface temperatures. See https://climate.nasa.gov/evidence/.
76. The probability that human activity is the root cause of climate change is greater than 95%. See https://climate.nasa.gov/evidence/
77. Such as how much such we emit by burning fossil fuels when running vehicles and powerplants, keeping livestock that release methane gas, and continuing deforestation of trees that absorb carbon dioxide.
78. Intergovernmental Panel Climate Change (IPCC) and NASA, *Climate Change 2014 Synthesis Report Summary for Policymakers*, based on IPCC, *Climate Change 2014: Synthesis Report. Contribution of Working Groups I, II and III to the Fifth Assessment Report of the Intergovernmental Panel on Climate Change* [Core Writing Team, R.K. Pachauri and L.A. Meyer (eds.)]. (Geneva: IPCC, 2014).
79. This includes projects that have been withdrawn due to opposition. See Lulu

and 2015, the City of San Francisco alone approved nearly 50 residential, retail, and office developments in locations fewer than eight feet above sea level,[80] even though real estate less than 10 feet higher than sea-water level is at risk for recurring or complete flooding by the year 2100. This, despite the fact that shortly before his death, Ed Lee, the mayor of San Francisco, stated that the city would standardize prediction of a 66-inch rise in sea water over the next five decades. Commerce seems to have outweighed protecting citizens.

The city also faces a major crisis with an existing sea wall that, if ruptured during a severe earthquake, could result in the flooding of most of the city's downtown, including the financial district. Yet despite the potential for such an enormous catastrophe and just as we don't prepare for the unintended consequences of exponential technological development, neither the government nor private investment is taking notice of this threat. Climate change will have a dramatic impact on businesses and governments. For example, insurance companies will have to change dramatically the way they price and evaluate risk, and governments will have to find new ways to finance infrastructure, such as sea walls.

Orozco, "Interactive Map: A Baywide Building Boom Threatened by Rising Waters," *San Francisco Public Press*, July 29, 2015.

80. Kevin Stark, Winnifred Bird, and Michael Stoll, "Major S.F. Bayfront Developments Advance Despite Sea Rise Warnings," *San Francisco Public Press*, July 29, 2015.

TOPIC X.

HUMANS AND WORK

We are currently experiencing an unprecedented pace of technological advancement. This will have a profound impact on labor markets, with many administrative jobs at risk for replacement by machines and cloud agents.

The consulting firm McKinsey estimates that upwards of 800 million jobs will be at risk by 2030 as a result of robotics and automation.[81] The advent of autonomous vehicles will put professional drivers out of work, and warehouses that are run by a large staff handling forklifts will be run by a few people overseeing fully automated robots. Fast-food orders can be made on touch screens, and a robot that makes hamburgers without any person ever touching the meat already exists.[82]

Increasingly complex tasks are being automated, resulting in improved speed, higher quality, and lower production costs for goods

---

81. James Manyika et al. (2017), "What the Future of Work Will Mean for Jobs, Skills, and Wages," McKinsey Global Institute, 2017. Also, James Manyika et al. (2017), "Harnessing Automation for a Future That Works," McKinsey Global Institute, 2017.
82. Kevin Delaney, "The Robot That Takes Your Job Should Pay Taxes, says Bill Gates," *Quartz*, February 17, 2017.

and services. On a positive note, AI will allow us to spend our time more efficiently, with technology performing routine administrative tasks, such as taxes and bookkeeping, so that humans can spend more time on activities we find more important or pleasurable.

But this transformation also raises the question of what is to happen to those who are no longer "useful" because they have been outperformed by machines. The U.S. has already experienced a dramatic increase in the number of workers who have left the workforce permanently, having given up on any prospect of finding a job. Although unemployment numbers are at a record low, these workers are not included in the numbers.

Even those workers who don't lose their jobs are likely to experience downward pressure on salaries as the number of unskilled jobs become comparably fewer in relation to the pool of job seekers. People competing for day-to-day employment at the U.K. Royal Job Exchange, as depicted in the novel, is just an extrapolation of current trends moving away from long-term employment.

Sharing services illustrate this development well. Where cab drivers used to have secure employment, they are now Uber or Lyft contractors who must bear the full consequences of changes in revenue and costs. Similarly, renting a room through Airbnb is a substitute for staying at a hotel, where concierges, maids, and other staff enjoy a more secure form of employment. On a higher level, firms are increasingly outsourcing services that are beyond their core business, such as a construction company that outsources invoicing and legal services.[83]

In many sectors, the outsourcing of services happens at the individual level. For example, news outlets often hire journalists on a contractual basis, and universities increasingly hire people with doctorate degrees for specific research projects or teaching assignments, rather than providing secure jobs that lead to tenure.

Besides the move away from employment security, several trends are concurrent and intertwined with the rapid rise of technology, most notably inequality and urbanization. Much of society's current transformation attributed to technology is happening in cities. Substantial

83. Pat McArdle et al., "Outsourcing Comes of Age: The Rise of Collaborative Partnering," PricewaterhouseCoopers, 2008.

gaps in employment and wages between urban and nonurban areas already exist, as highly educated households are concentrated in cities.[84] Because of agglomeration economics, due to the value of the concentration of human capital, people make more money in cities regardless of their level of competence. The concurrent trends of automation, specialization, urbanization, and the rise of technology are likely to exacerbate the income gap between low- and high-skilled individuals.

Government policy, most notably regarding taxation, will have to change as business owners will need fewer workers, increasing profits and lowering overall tax revenue. In a 2017 interview, Bill Gates proposed a robot tax.[85] When a robot replaces a factory worker who makes, say, $50,000, Gates proposes that a similar level of taxation (income tax and Social Security payments) should be imposed on the employing company to temporarily slow the spread of automation and to finance other jobs, such as taking care of the elderly or teaching, which are believed to be better performed by humans.

Gates believes that slowing down the pace of automation would allow governments to figure out a way to handle a not-so-distant future when most jobs are replaced by machines. Personally, I don't believe the pace of technology advancement can be slowed. The outsized influence of corporate donors on Congress will prevent legislation to artificially slow the speed of innovation.

The novel outlines a society that is far more divided and unequal than what we see today. Current trends suggest we are headed in this direction. Decision makers will need to address these trends, or greater cultural division and economic inequities may become yet another unintended consequence of progress.

---

84. A large body of research supports a positive relationship between city size and growth in employment and population. Studies also support a positive relationship between city size and wages. See Donner, Eriksson, and Steep, "Digital Cities: Real Estate Development Driven by Big Data." (rest of the citation?)
85. Delaney, "The Robot That Takes Your Job Should Pay Taxes."

FIRST LIGHT OF DAY DISCUSSION QUESTIONS

1. In the book, society has evolved where both the positive and negative consequences of this are readily apparent to the reader but not to the characters themselves. Should we think about disruptive technology differently by looking at the social consequences of technology products and services?

2. How will today's unprecedented explosion in global technology innovation impact what it means to be a citizen functioning in a society?

3. In Silicon Valley and the San Francisco Bay Area we are seeing the vast divergence in wealth create a geographical area that is unaffordable to most of the area's workers. Should this be mitigated, and if so, by government or industry?

4. Can and should machines understand ethics, compassion, and morality? Would sentient beings have a soul?

5. How should the explosion in disruptive technology influence what we teach in schools?

6. How (and what) should humanity do to protect itself from the unintended consequences of technology?

7. How should we think about the impact disruptive technology the economy of the future? Is it possible to transform disruption into new opportunity for growth?

8. We talk about unemployment being at an all-time low. But what about the number of people who have left the workforce—and who are no longer being counted in the unemployment numbers— now at an all-time high?

9. Are we beginning to see the emergence of a zero-value population that no longer is capable of producing value for society?

10. If you were in Mikhail's position, what decision would you make that could change the future of humanity?

ACKNOWLEDGMENTS

I wish to express my deep appreciation to Ray Levitt, emeritus professor at the Stanford Engineering School and co-founder of the Disruptive Technology and Digital Cities Program, for the inspiration to write this book. Many thanks to John Hennessy, former president of Stanford University, for taking the time out of his busy schedule to offer advice on publishing.

I would also like to thank Christine Walker, my lifelong companion and wife, for her enduring support of a maverick's behavior that led to this adventure.

I would like to thank Dr. Herman Donner, my willing accomplice, researcher, and writer, for his contribution to the companion guide.

Finally, thanks and acknowledgment to the entire Silicon Valley Press team, including my friend and mentor Joseph DiNucci, and Cheryl Dumesnil and Atiya Dwyer, for making this book possible.

ABOUT THE AUTHOR

Michael Steep has been at the forefront of technology for the last 30 years. He is the founder and director of Stanford University's Disruptive Technology & Digital Cities Program, and a frequent speaker on leading innovation. His expertise on disruption and innovation comes from the field as senior vice president of Xerox PARC and earlier in executive and management positions at leading tech giants including Hewlett-Packard, IBM, Microsoft, and Apple. He resides with his wife in Silicon Valley.